About Pearson

Pearson is the world's learning company, with presence across 70 countries worldwide. Our unique insights and world-class expertise comes from a long history of working closely with renowned teachers, authors and thought leaders, as a result of which, we have emerged as the preferred choice for millions of teachers and learners across the world.

We believe learning opens up opportunities, creates fulfilling careers and hence better lives. We hence collaborate with the best of minds to deliver you class-leading products, spread across the Higher Education and Test Preparation spectrum.

Superior learning experience and improved outcomes are at the heart of everything we do. This product is the result of one such effort.

Your feedback plays a critical role in the evolution of our products and you can contact us - reachus@pearson.com. We look forward to it.

AI for Everyone

A Beginner's Handbook for Artificial Intelligence

Saptarsi Goswami

Assistant Professor and Head of the Department
Department of Computer Science
Bangabasi Morning College affiliated to University of Calcutta

Amit Kumar Das

Principal Director
LTIMindtree

Amlan Chakrabarti

Professor and Director
A.K. Choudhury School of Information Technology
University of Calcutta

Senior Manager—Product: Neha Goomer
Senior Editor—Production: C. Purushothaman

ISBN 978-93-615-9175-4

First Impression, 2024
Sixth Impression, 2025
Seventh Impression, 2026

Published by Pearson India Education Services Pvt. Ltd, CIN: U72200TN2005PTC057128.

Head Office: 1st Floor, Berger Tower, Plot No. C-001A/2, Sector 16B, Noida - 201 301, Uttar Pradesh, India.
Registered Office: Featherlite, 'The Address' 5th Floor, Survey No 203/10B, 200 Ft MMRD Road, Zamin Pallavaram, Chennai – 600 044.
Website: in.pearson.com, Email: companysecretary.india@pearson.com

Compositor: SRS Global, Puducherry
Printer at Rajkamal Electric Press, Kundli, Haryana-131 028

Contents

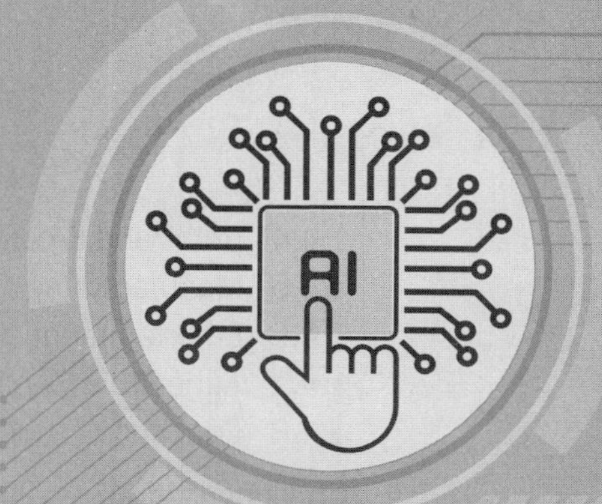

Chapter 3   Industrial Applications of AI   75

Chapter 5   AI in Research, Generative AI and Other Important Issues   137

Foreword

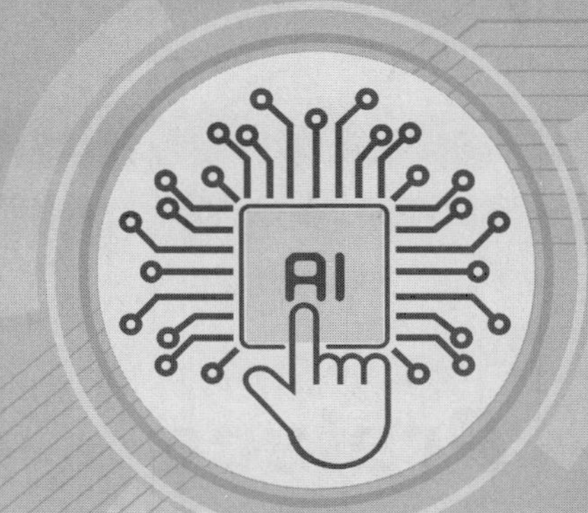

The core idea of man-made objects demonstrating human-like intelligence dates back to ancient myths and is part of folklore. Ever since the term "Artificial Intelligence" was coined in 1956, this subject has attracted immense interest from within the scientific community, while also captivating the imaginations within science-fiction writers or movie-makers. Across the last two decades, the tremendous advancement of computing power, algorithms and availability of data has led to further development of AI as a subject. Over the years, AI has undergone periods of hype and skepticism, but there is absolutely no doubt that it will continue to evolve and influence diverse fields, promising both opportunities and challenges for the future.

Is AI going to make our lives easier? What are the associated security challenges? Is it ethical to disclose data? What are the social implications? These are some of the very valid points one must be aware of, irrespective of one's background and area of operation. Healthcare, manufacturing, finance, education, retail – are just some of the sectors that are gearing up to adopt AI in various forms – mainly to boost productivity.

Overall, the future of mankind with AI is likely to be characterized by increased efficiency, innovation, and convenience, but also by the need for thoughtful regulation, ethical considerations, and societal adaptation. This book is perfectly timed and aims to create awareness and provide clarity around this subject. The content of the book is perfectly suited to impart sufficient knowledge to students and other readers with diverse backgrounds, so that they have basic familiarity with the key concepts in AI.

Sanjay Guha Thakurta,
Executive Partner, IBM India.

Preface

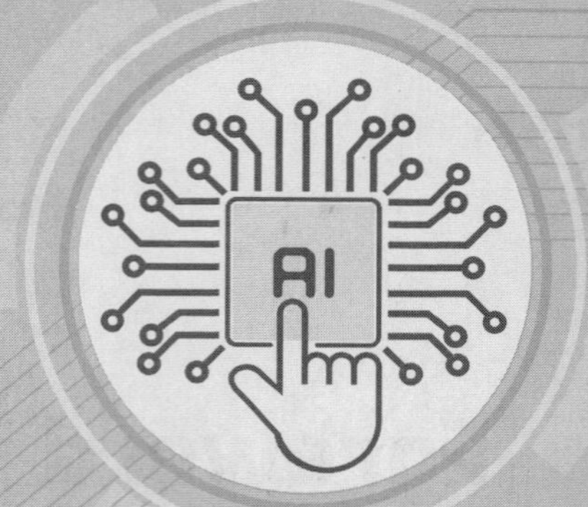

In the modern digitized business world, Artificial Intelligence (AI) and it's applications have become quite imperative - so much so that from healthcare to manufacturing to education, applications of AI in some form or other is inevitable. With such widespread adoption of AI cutting across different business domains as well as walks of life, it is extremely important for "Everyone" to understand at least the basic concepts of AI.

AI has been given due importance by the University Grants Commission (UGC) while formulating the *Curriculum and Credit Framework for Undergraduate Programmes* based on the recommendations of *National Education Policy (NEP), 2020.* The NEP 2020, brought in by Ministry of Human Resource Development, Government of India, recognizes that higher education plays an extremely important role in promoting human as well as societal well-being and in developing India. It focuses on preparing professionals in cutting-edge areas that are fast gaining prominence, such as AI. Hence AI should be introduced to the wide spread of students in all streams namely Science, Commerce and Humanities. This book is a humble attempt to introduce a learner to the area of Artificial Intelligence.

The thought of authoring this book title came to our mind based on an all-round survey of teaching resources available to introduce the subject of Artificial Intelligence or AI to the wide mass of students. While authoring this book, due attention has been given to the fact that the content of the book needs to be lucid enough to be readable by students from the different faculties. At the same time, it has been taken care of that the coverage is comprehensive. This book has been written in simple English and to present the AI concepts in an easily understandable way which can be used as a textbook for both graduate students as well as general readers.

About the Book

Readers of this book will gain a basic understanding of AI concepts. Not only students, but also the general readers will find a variety of concepts related to AI. Students, software developers, technical managers as well as general readers with no background in computer science or programming will find the material in this book easily readable.

Style of Writing

- The language used in the book is lucid, is easy to understand, and facilitates easy grasping of concepts.
- The chapters have been logically arranged in sequence.
- Each chapter starts with an introductory paragraph which gives overview of the chapter.

- Discussion points and Point to Ponder given inside the chapters help to clarify and understand the chapters easily and explores the thinking ability of the readers.
- Sample questions, at the end of each chapter, helps the students to prepare for the examination. Objective type questions, short questions and long questions are given which helps students to preparing the concepts easily for the exam.
- Most of the content presented in the book is in the form of bullets, organized sequentially. This form of presentation, rather than in a paragraph form, facilitates the reader to view, understand and remember the points better.
- The explanation is supported by diagrams, pictures and images wherever required.

How This Book is Organized?

- Several latest topics have been included in the book. Some of these topics are:
  - The concept of Generative AI and how it is related with NLP
  - The perspectives of Ethical AI
  - Latest industry applications of AI
- The book starts with a simple introduction to the concept of artificial intelligence. Therein, it also covers a complete anecdote of how AI has evolved through the past few decades.
- Then it delves into different sub-fields and technologies under AI.
- After that it covers in details the application use cases in some major domains like healthcare, finance, retail, etc.
- Then it discusses the different ethical aspects of AI like bias and fairness, transparency, inclusivity, etc.
- At the end, some highlight is given on the research aspects of AI and some key upcoming trends like GenAI.

Target Audience

This book has been written keeping in mind the readers familiar and not-so familiar with computers. Several chapters have been included that cover the syllabi of different universities in India. The book is well suited for the following target audience:

- Computer science students undergoing a course in computer science—DCA, MCA, BCA, DOEACC level courses.
- Engineering students of first year—BTech, BE.
- Science students pursuing BSc in Physics, Chemistry, Botany, Zoology and Mathematics.
- Non-science students pursuing BCom (P), BCom (Hons), BA (P), BA (Hons), BDP, BBA, MBA, BBE
- Students enrolled in short-term courses on IT in polytechnics, training institutes, Technical Institutes.
- Management students pursuing BBA, MBA, PGDBM
- Any learner interested

Model Syllabus for AI for Everyone

Credits: 3	Contacts per week: 3 lectures

Chapter 1 - Introduction to Artificial Intelligence: Definition and scope of AI, Historical overview and key milestones, Differentiating AI from human intelligence.

[6 Lectures]

Chapter 2 - AI Subfields and Technologies: Machine learning: Supervised, unsupervised, and reinforcement learning, Deep learning and neural networks, Natural language processing (NLP) and computer vision.

[10 Lectures]

Chapter 3 - Applications of AI: AI in healthcare; AI in finance; AI in Retail; AI in Agriculture; AI in education; AI in transportation.

[12 Lectures]

Chapter 4 - Bias and Fairness in AI Systems: Privacy and data protection concerns; Impact of AI on employment and the workforce, AI and social inequality

[6 Lectures]

Chapter 5 - AI in Research, Generative AI and other important issues: AI and Innovation: Emerging trends and future directions in AI

[6 Lectures]

Lesson Plan

Chapter #	Chapter Name	Detailed content	Lectures allotted
1	Introduction to Artificial Intelligence	What is AI?	Lecture 1, 2 and 3
		The journey of AI	
		Artificial General Intelligence (AGI)	Lecture 4 and 5
		Brief on industry applications of AI	
		Challenges of AI	Lecture 6
2	AI Subfields and Technologies	Brief introduction to AI sub-fields	Lecture 7
		Computer Vision	Lecture 8
		Natural Language Processing	Lecture 9
		Machine Learning	Lecture 10 and 11
		Deep Learning	Lecture 12
		Reinforcement Learning	Lecture 13
		Robotics	Lecture 14
		Knowledge Engineering	Lecture 15 and 16
3	Applications of AI	In Healthcare	Lecture 17 and 18
		In Finance	Lecture 19 and 20
		In Retail	Lecture 21 and 22
		In Agriculture	Lecture 23 and 24
		In Education	Lecture 25 and 26
		In Transportation	Lecture 27 and 28
4	Bias and Fairness in AI Systems	Introduction	Lecture 29
		Ethics in AI	Lecture 30
		Bias and Fairness in AI systems	Lecture 31
		Transparency in AI systems	Lecture 32
		Accountability of AI systems	
		Privacy and Data protection concerns	Lecture 33
		Security of AI models	
		Inclusivity in AI	Lecture 34
		Sustainability in AI	
		Robustness and Reliability	

Chapter #	Chapter Name	Detailed content	Lectures allotted
5	AI in Research, Generative AI and other important issues	AI in Research and Experimentation	Lecture 35
		Generative AI	Lecture 36, 37 and 38
		Synopsis of the Important Models	
		ChatGPT and Prompt Engineering	
		Emerging Trends in AI	Lecture 39
		AI and Social Equality	Lecture 40

Acknowledgements

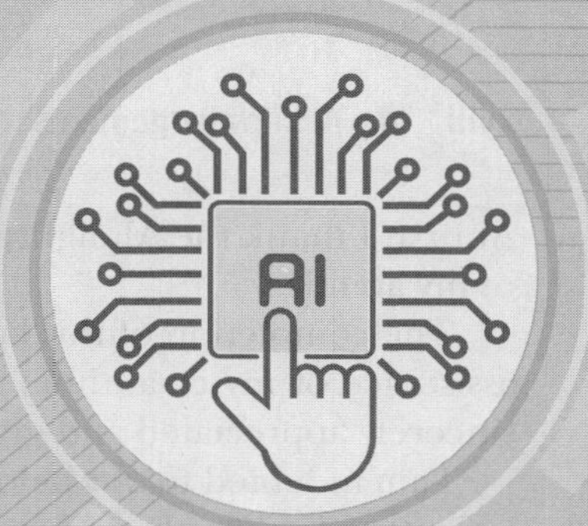

We are grateful to Pearson Education, who accepted our proposal came forward to publish this book. We are also indebted to Ms. Neha Goomer and Mr. Purushothaman Chandrasekaran of Pearson Education for their all-out support and collaboration throughout the journey. Their patience and guidance were invaluable. Thank you very much Neha and Purush.

Saptarsi, Amit and Amlan

I would start by offering my respect to Debi Saraswati. "*Ya Devi sarvabhuteshu, Vidya rupen samasthita, Namastasye Namastasye Namastasye Namo Nmamah.*"

This is my second book with Pearson Education. Thanks to Pearson Education for having their trust upon me. I thank my co-authors Dr. Amit Kumar Das, Prof. Amlan Chakrabarti without whose active contribution this book would not have been possible. I thank my students, research scholars, my colleagues, fellow members of this knowledge community, my college Bangabasi Morning for the constant inspiration they provide on daily basis.

I must acknowledge my family for their unwavering support and understanding during the time I dedicated to writing this book. I am indebted to my father, Prof. Amar Sankar Goswami, and my mother, Dr. Swati Goswami, for their boundless blessings and affection. Special thanks to my life partner, my wife Sanchari, and my sons Udaybhan and Ujaan, whose unconditional love and encouragement have been my constant companions throughout this journey.

Saptarsi Goswami

First, I would like to thank the Almighty for getting the opportunity to be a part of this project. Next, I would like to thank my family for the constant support and encouragement. My parents have always been my role model, my wife a source of constant strength and my daughters are my happiness. Without whole-hearted support from my family, I would not have got the luxury of time to spend on hours in writing the book. Any amount of thanks will fell short to express my gratitude and pleasure of being part of the family.

This is the fourth book that I have written with Pearson Education. Thanks a lot Pearson Education for keeping trust on me – means a lot as an author. Special thanks to Neha and Purush for providing constant support and time.

Amit Kumar Das

At first, I thank the Almighty for providing me the strength to carry out the work at the best of my abilities.

The completion of this book could not have been possible without the participation and assistance of my co-authors and support from Pearson Education. Their contributions are sincerely appreciated and gratefully acknowledged.

I am indebted to our beloved Professor Late Prof. Subhansu Bandopadhyay, University of Calcutta who has always inspired me in all my academic achievements. I'm also immensely grateful to Prof. Susmita Sur-Kolay, Indian Statistical Institute, Kolkata, who my Ph.D. supervisor and a constant source of support and guidance in all of my academic endeavors.

I thank my research scholars who have always been my core strength and helping this book project through their valuable inputs, that have emerged during the various in-depth discussions on the subject matter of this book.

I would like to thank my father Amalendu Chakrabarti, whose love and guidance are always with me and is a true source of blessings for me. Most importantly, I wish to thank my loving and supportive wife, Sarani, and my two wonderful children, Sambuddha and Aniruddha who provide unending motivation for me.

Amlan Chakrabarti

Pearson would like to thank following experts who have reviewed the manuscript and provided their valuable inputs and suggestions:

- **Anirban Ray**, Bangabasi Morning College, Kolkata, West Bengal

- **M. Uma Priya**, Hindusthan College of Engineering & Technology, Coimbatore, Tamil Nadu

- **Mousoomi Bora**, The Assam Kaziranga University, Assam

- **Tanmoy Biswas**, Syamaprasad College, Kolkata, West Bengal

About the Authors

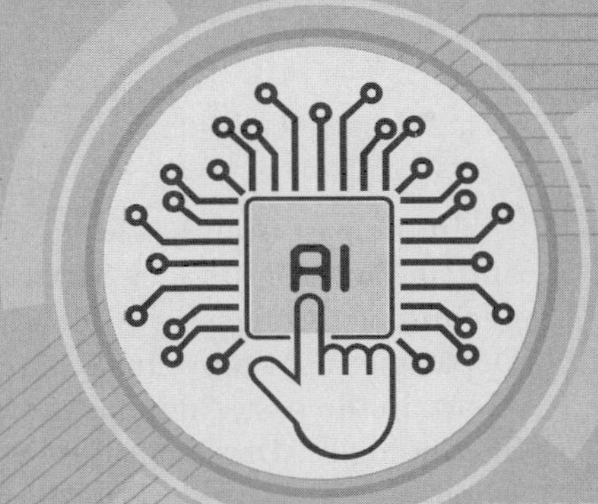

Dr. Saptarsi Goswami is currently working as an Assistant Professor and Head of the Department – Computer Science, Bangabasi Morning College affiliated to University of Calcutta. He has a total of 22 years of experience – 11 years in academics and 11 years in IT Industry in companies like PwC, Cognizant and Tata Infotech. He has done his B.Tech. from NIT, Jaipur and his M.Tech. and PhD from University of Calcutta. He is a visiting faculty and research supervisor in A.K. Choudhury School of IT University of Calcutta and has been a founding member of the Data Science Lab, University of Calcutta.

Saptarsi has more than 75 publications in reputed international journals and conferences with an H-Index of 19. He is deeply involved in many community initiatives in Data Science, Machine Learning and Artificial Intelligence. His YouTube lectures on AI and Machine Learning are part of National Digital Library, India and was adjudged Person of the Week by NDL in July 2020. He is an executive member of Society for Data Science (S4DS). He regularly speaks at different industry and academic forums and writes at towards datascience. He has supported Higher Education Department, Govt. of WB for various IT Projects, as well as capacity building initiatives in an advisory role.

Dr. Amit Kumar Das has been working in the IT industry and in academic research/teaching in Data Science for the past 25 years. He is currently working Principal Director at LTIMindtree. Amit has done Ph.D. from University of Calcutta and M.Tech. from Birla Institute of Technology & Science, Pilani. Before that, he did graduation in Electrical Engineering from Bengal Engineering College, Shibpur (currently known as IIEST).

Amit has many research papers in Data Analytics and Machine Learning published in international journals and conferences. He has co-authored textbooks on Machine Learning, Deep Learning, and Big Data published by Pearson, primarily for the under-graduate students. He also has been a regular speaker in Data Analytics and Machine Learning. He is a Senior Member of IEEE.

In the past Amit has worked at Cognizant Technology Solutions for more than a decade and at Tata Consultancy Services. Before joining his current role, he was working as Chief Technology Officer in a tech startup. He has also spent few years as a faculty member, Department of Computer Science and Engineering, Institute of Engineering & Management. Apart, in the past, Amit has provided consulting in Data Science with companies like Rebaca Technologies, Dirac Business Solutions Pvt. Ltd and IT program management support to Department of Higher Education, Government of West Bengal.

Prof. Amlan Chakrabarti is a Full Professor in the A.K. Choudhury School of Information Technology at the University of Calcutta and Adjunct Professor at IIIT Delhi. He was a post-doctoral fellow at the School of Engineering, Princeton University, USA during 2011–2012. He has almost 20 years of experience in Engineering Education and Research. He is the recipient of DST BOYSCAST fellowship award in Engineering Science (2011), Indian National Science Academy (INSA) Visiting Faculty Fellowship (2014), JSPS Invitation Research Award (2016), Erasmus Mundus Leaders Award (2017), Hamied Visiting  Professorship from University of Cambridge, UK (2018) and prestigious *Siksha Ratna* Award by Dept. of Higher Education Govt. of West Bengal (2018). He has also served in various capacities in various higher education organizations both at national and international levels.

Amlan has published around 200+ research papers in referred journals and conferences and has graduated 20+ Ph.D. students. He is the Series Editor of Springer Transactions on Computer Systems and Networking, Associated Editor of the Elsevier Journal of Computers and Electrical Engineering and Guest Editor of the Springer Journal of Applied Sciences. He is a Senior Member of IEEE and ACM, IEEE Computer Society Distinguished Visitor, Distinguished Speaker of ACM, Vice Chair of IEEE CEDA India Chapter, Vice President of Data Science Society and Life Member of CSI India. His areas of research are Machine Learning, Computer Vision, Reconfigurable Computing, Cyber-physical Systems, VLSI CAD and Quantum Computing.

Introduction

Deep in the annals of scientific history, in the early 1950s, a group of brilliant and visionary scientists embarked on a crucial journey that would forever alter the course of technological advancement. Led by the legendary computer scientist John McCarthy, this exceptional group explored the uncharted territory of creating an entity that could replicate and even surpass human intelligence—this concept would later be coined as artificial intelligence (AI).

Figure 1.1 Large computing machines

During this era, computers were colossal machines (refer Fig. 1.1), but possessed limited processing power. The notion of constructing a device capable of thought, reasoning, and understanding seemed audacious and perhaps unattainable. Nonetheless, McCarthy and his committed team undeterred by the enormity of the challenge and fuelled by a never-ending curiosity to push the boundaries of human achievement.

The birth of AI as a formal area can be traced back to a historical event in 1956—the Dartmouth Conference. McCarthy, along with fellow luminaries in the field, including Marvin Minsky, Nathaniel Rochester and Claude Shannon, orchestrated this historic gathering that convened a remarkable assemblage of prominent scientists, mathematicians and intellectuals, all united by a shared fascination with the prospect of creating intelligent machines. John McCarthy developed LISP (List Processing) in 1958. LISP became a fundamental programming language for AI research and remains influential in AI development.

During the Dartmouth Conference, the participants engaged in spirited discussions, contemplating the intricate facets of AI. Conversations reverberated with the exploration of topics like natural language processing, problem-solving, perception and machine learning. It was at this conference that the term 'artificial intelligence' was first uttered—a significant moment that heralded the formal emergence of a groundbreaking scientific discipline. In the years that followed, AI research progressed. Determined researchers diligently honed their skills, grappling with the fundamental challenges of creating intelligent machines. Algorithms were conceived, programming techniques devised and conceptual frameworks established to simulate the intricate nuances of human thought processes.

POINTS TO PONDER

John McCarthy introduced the term 'artificial intelligence' in 1956 during the Dartmouth Conference, which marked the birth of AI as a formal field of study. The impact of the Dartmouth Conference can be measured by the exponential growth of AI research and applications in subsequent decades. It catalysed the development of AI as a recognized discipline and paved the way for future advancements.

1.1 THE JOURNEY OF ARTIFICIAL INTELLIGENCE (AI)

1.1.1 The Early Years of AI: 1940s and 1950s

The 1940s and 1950s were formative years for AI research, characterized by foundational work in computer science. During this period, AI research received significant support from government agencies, particularly in the United States. Organizations like the DARPA (Defense Advanced Research Projects Agency) funded research projects in AI, recognizing its potential for military and civilian applications. The roots of AI can be traced back to World War II when researchers were working on projects related to automatic computation and information processing. During the war, scientists like Alan Turing in the UK and John von Neumann in the US developed the theoretical foundations of computing machines, which later became essential components of AI research.

In 1943, McCulloch, a neurophysiologist, and Pitts, a logician, collaborated to create a mathematical model of a simplified artificial neuron, often referred to as the 'McCulloch-Pitts neuron' or simply the 'M-P neuron.' This model aimed to mimic the basic functioning of a biological neuron. The McCulloch-Pitts neuron was a foundational concept in the early development of artificial intelligence and neural network theory. It provided a mathematical framework for modelling basic decision-making processes inspired by biological neurons. While it was a simplified and abstract representation of real neurons, it was a starting point for further research into more complex neural network architectures. Their work influenced later researchers like Frank Rosenblatt, who developed the 'perceptron' (the first shallow neural network) in 1957. It ultimately contributed to the modern field of deep learning, which has become a cornerstone of contemporary artificial intelligence.

1.1.2 The Initial Journey: 1960s

The 1960s saw early attempts at natural language processing (NLP). Researchers like Joseph Weizenbaum developed programs like ELIZA, which could engage in text-based conversations with users, albeit in a limited way. In the late 1960s, Stanford Research Institute (now SRI International) developed Shakey, one of the first mobile robots with sensors to navigate its environment. While its capabilities were limited, it laid the groundwork for robotics and AI integration. The 1960s also saw significant progress in symbolic AI, emphasizing the use of symbols and logical reasoning. Researchers developed methods for representing and manipulating knowledge, paving the way for expert systems and knowledge-based AI. Alexey Grigoryevich Ivakhnenko, a Soviet mathematician and computer scientist, considered to be the 'Father of Deep Learning' developed the first multi-layer perceptron in 1965, which forms the basis of deep learning.

However, by the end of the 1960s, the field of AI confronted formidable obstacles that led to a tumultuous period. The initial enthusiasm and optimism began to wane as early AI systems could not live up to the high expectations set by researchers and the public alike. The nascent field grappled with the inherent limitations of computing power, the scarcity of extensive datasets and the complexity of human cognition. In 1969, the researchers Marvin Minsky and Seymour Papert published a book '*Perceptron*' pointing to the fact that neural networks have failed to meet expectations.

1.1.3 First AI Winter: 1970s

The first AI winter occurred in the early 1970s. Initial enthusiasm for AI research led to significant investment and ambitious goals. DARPA funded many AI research projects. However, progress in AI did not match expectations, and the limitations of computing technology became apparent. Funding for AI research decreased, and many AI projects were discontinued. DARPA started to cut funds in 1970 due to a lack of enthusiasm.

The famous Lighthill Report, compiled by researcher James Lighthill commissioned by the British government in 1973, criticized AI research for its perceived lack of progress and questioned the allocation of funding. This report further contributed to the scepticism surrounding AI research during the decade. However, the 1970s did not signal the end of AI research. Instead, it provided valuable lessons. Researchers learned that AI technology was not ready to achieve the lofty goals set forth. Yet, they remained determined and resilient. The 1970s were a pivotal period that prompted AI researchers to re-evaluate their approaches and expectations. It was clear that advances in computing technology were essential for AI to fulfil its potential.

1.1.4 Renewed AI Excitement: 1980s

The 1980s represented a recovery period following the AI winter of the 1970s. Researchers learned from past challenges and demonstrated that AI could deliver practical solutions, especially in knowledge-based systems. The 1980s saw an increase in the development and adoption of expert systems across various industries. These systems demonstrated expertise in tasks ranging from medical diagnosis to financial planning. Examples include medicine and manufacturing.

In 1983, DARPA funded the Connection Machine project, led by Danny Hillis. The project aimed to develop a massively parallel supercomputer. In the mid-1980s, Japan launched the ambitious Fifth Generation Computer Systems Project. The goal was to develop advanced computer systems with AI capabilities. While the project faced challenges and ultimately did not achieve all its objectives, it underscored Japan's commitment to AI research.

The 1980s also saw the development of specialized computers, called LISP machines, optimized for running LISP, a programming language commonly used in AI research. While these machines were designed to boost AI development, their market was limited, and they faced challenges in gaining widespread adoption. Towards the late 1980s, neural networks began to regain attention. Researchers like Geoffrey Hinton and Yann LeCun made significant contributions to developing neural network algorithms, paving the way for the resurgence of machine learning in the following decades.

By the end of the 1980s, the second AI winter followed a period of overinflated expectations in the 1980s. The field of expert systems, which aimed to replicate human expertise in specific domains, faced challenges in scaling up and delivering practical applications. Funding declined and AI research could not find real-world applications. In 1987, DARPA reduced the funding for AI research again. The market for LISP machines also collapsed.

1.1.5 AI in the 1990s

The 1990s represented a dynamic era of Artificial Intelligence (AI). This decade witnessed significant progress in AI research, as well as some persistent challenges. The 1990s marked a resurgence of interest in machine learning, a subfield of AI focused on creating algorithms that could learn from data. Researchers began developing more sophisticated machine learning techniques, including neural networks and decision trees. Reinforcement learning, a type of machine learning where agents learn to make decisions through trial and error, gained prominence during the 1990s. Researchers explored its applications in robotics and game playing.

In 1997, IBM's Deep Blue supercomputer famously defeated world chess champion Garry Kasparov, as can be seen in Fig. 1.2. Deep Blue had a specialized chess chip and a sophisticated search algorithm. Garry Kasparov, on the other hand, was one of the greatest chess players in history, holding the World Chess Champion title since 1985. The first official match between Kasparov and Deep Blue took place in 1996. Deep Blue won the first game, marking a significant moment in AI history. However, Kasparov went on to win the match 4-2. The highly anticipated rematch occurred in May 1997 in New York City. It was a best-of-six-game match. Deep Blue has significantly improved its hardware and software since the previous encounter. In a shocking turn of events, Deep Blue won the match against Kasparov by a score of 3.5 to 2.5, becoming the first computer program to defeat a world chess champion in a match with standard time controls. Deep Blue's victory was seen as a significant milestone in AI research, demonstrating the potential of computers to excel in complex intellectual tasks. It sparked discussions about the relationship between human intelligence and artificial intelligence.

Figure 1.2 Garry Kasparov faced off against Deep Blue

POINTS TO PONDER

- Deep Blue was able to imagine an average of 200,000,000 positions per second.
- Despite of his loss to IBM Deep Blue, Garry Kasparov remained a strong advocate for human-computer collaboration in chess. He later played in 'Advanced Chess' tournaments, where both humans and computers teamed up.

Robotics research made significant headway during the 1990s. Robots have become more capable of tasks like navigation and object manipulation. Research in autonomous vehicles also gained traction. The late 1990s saw a surge in AI-related startups, fuelled by the dot-com bubble. However, many of these startups struggled to deliver on their promises, leading to scepticism and a subsequent burst of the dot-com bubble.

1.1.6 The Reinvention of AI: The 2000s Revolution

The first decade of the 21st century have marked a pivotal period of artificial intelligence (AI) research, with several vital incidents reshaping the field and propelling it into the mainstream. These incidents collectively brought AI into various facets of daily life, reshaping how we interact with technology and information.

The early 2000s witnessed the emergence of online platforms such as YouTube, social media, and e-commerce websites. These platforms generated an unprecedented amount of data, offering researchers a goldmine of information for AI development. Researchers capitalized on this data deluge to create algorithms for video analysis, content recommendation and sentiment analysis. The ability of AI to harness vast amounts of data transformed how businesses operated and how people consumed content.

Another critical incident was the advancement of reinforcement learning techniques. AI researchers like Andrew Ng and Stuart Russell significantly contributed to this domain. Reinforcement learning allows AI systems to learn by interacting with their environments. In 2006, an AI program made headlines by mastering the game of chess, demonstrating the potential for AI to learn complex strategies independently.

Throughout the 2000s, researchers achieved significant breakthroughs in computer vision. These advancements laid the foundation for AI applications in facial recognition, autonomous vehicles, and image analysis. As online shopping and content consumption surged, companies like Amazon and Netflix leveraged AI to develop sophisticated recommendation systems. These systems analysed user behaviour and preferences to provide personalized product and content recommendations. This innovation transformed how consumers discovered and engaged with products and media.

1.1.7 The AI Renaissance: Transforming the 2010s

The second decade of the 21st century have witnessed an unprecedented surge in AI research and development, characterized by groundbreaking incidents that reshaped industries and society. The 2010s marked the rise of deep learning, a subset of machine learning which is based on artificial neural networks. In 2012, a milestone incident occurred when a deep learning model named AlexNet won the ImageNet competition, significantly advancing image recognition accuracy. This event triggered a deep learning revolution, leading to remarkable progress in computer vision, speech recognition, and natural language processing.

In 2016, an AI program named AlphaGo, developed by DeepMind, defeated the world Go champion, Lee Sedol. This watershed moment showcased AI's capability to tackle complex, strategic games previously considered beyond the reach of machines. AlphaGo's victory demonstrated the power of deep reinforcement learning and set a new standard for AI's potential in solving intricate problems.

Autonomous vehicles became a focal point of AI research in the 2010s. Companies like Tesla, Waymo, and Uber initiated ambitious projects to develop self-driving cars. Notably, in 2015, Tesla introduced its Autopilot feature, enabling semi-autonomous driving on highways. This incident marked the beginning of a transformative era in transportation. The 2010s also witnessed significant strides in natural language processing (NLP). In 2018, OpenAI introduced GPT-2, a language model demonstrating unprecedented text generation capabilities. This innovation led to advancements in chatbots, virtual assistants, and language understanding, influencing communication and content generation across various domains.

DID YOU KNOW?

- *AI can beat humans in complex games like chess and Go?*
  DeepMind's AI program, AlphaZero, defeated world champions in chess and Go, showcasing AI's ability to master complex strategy games through machine learning.
- *AI can help farmers improve crop yields?*
  AI systems can monitor crop health, soil conditions, and weather data to optimize irrigation and fertilizer use, increasing agricultural productivity and sustainability.
- *AI can assist in drug discovery?*
  AI algorithms analyse vast chemical databases and predict how molecules will interact, accelerating the discovery of new medicines and treatments.

1.1.8 And the AI Journey is on …

Despite the setbacks, the indomitable spirit of AI research persevered, and the field experienced a renaissance in the 1980s and 1990s. Advances in computing technology, the growing availability of large datasets, and the refinement of algorithms breathed new life into the realm of AI. Expert systems, which captured the knowledge and reasoning of human experts in specific domains, gained widespread popularity during this period, ushering in a wave of practical AI applications.

At the dawn of the 21st century, AI encountered a paradigm-shifting transformation. The confluence of technological breakthroughs unleashed a wave of unprecedented progress. The advent of colossal computational power, coupled with the deluge of data inundating the digital landscape, propelled the development of groundbreaking algorithms—most notably, deep learning. Inspired by the intricate structure and functioning of the human brain, deep learning algorithms empower machines to acquire knowledge and insights from vast datasets, enabling them to make increasingly complicated decisions.

Presently, AI has transcended the realm of scientific speculation and firmly embedded itself in the tapestry of modern life. Its pervasive influence permeates diverse domains, fuelling the engines of innovation and redefining the boundaries of human potential. AI powers voice assistants, recommendation systems, autonomous vehicles, medical diagnosis tools and an array of other applications that have become indispensable to our daily lives.

As the relentless march of progress continues, AI stands at the forefront of technological advancement, propelling society toward a future characterized by unrivalled innovation. Reinforcement learning, natural language processing, computer vision, robotics—these are just a few realms where AI research continues to push the boundaries of possibility. With each passing day, scientists, engineers and visionaries unveil new frontiers, encouraged by the immense promise of AI.

The journey of AI, adorned with both triumphs and tribulations, stands as a testament to the indomitable human spirit and our insatiable thirst for knowledge. From the lofty aspirations and intellectual musings of visionaries to the realization of intelligent machines, the narrative of AI embodies the profound impact that human ingenuity can have on the world.

As we gaze into the boundless horizon of the future, it is abundantly clear that the dream of creating machines that can comprehend, learn and adapt is steadily metamorphosing into a tangible reality. As technology continues to progress at a rapid pace, the possibilities for AI remain limited only by the frontiers of our collective imagination. The future promises to be a tapestry woven with even more remarkable achievements, where the convergence of human intellect and artificial intelligence shall forge a path toward a world defined by unparalleled progress and infinite potential. Some of the key events in the evolution of deep learning have been captured in Fig. 1.3, though there were many more significant researches which happened over the years but couldn't be captured.

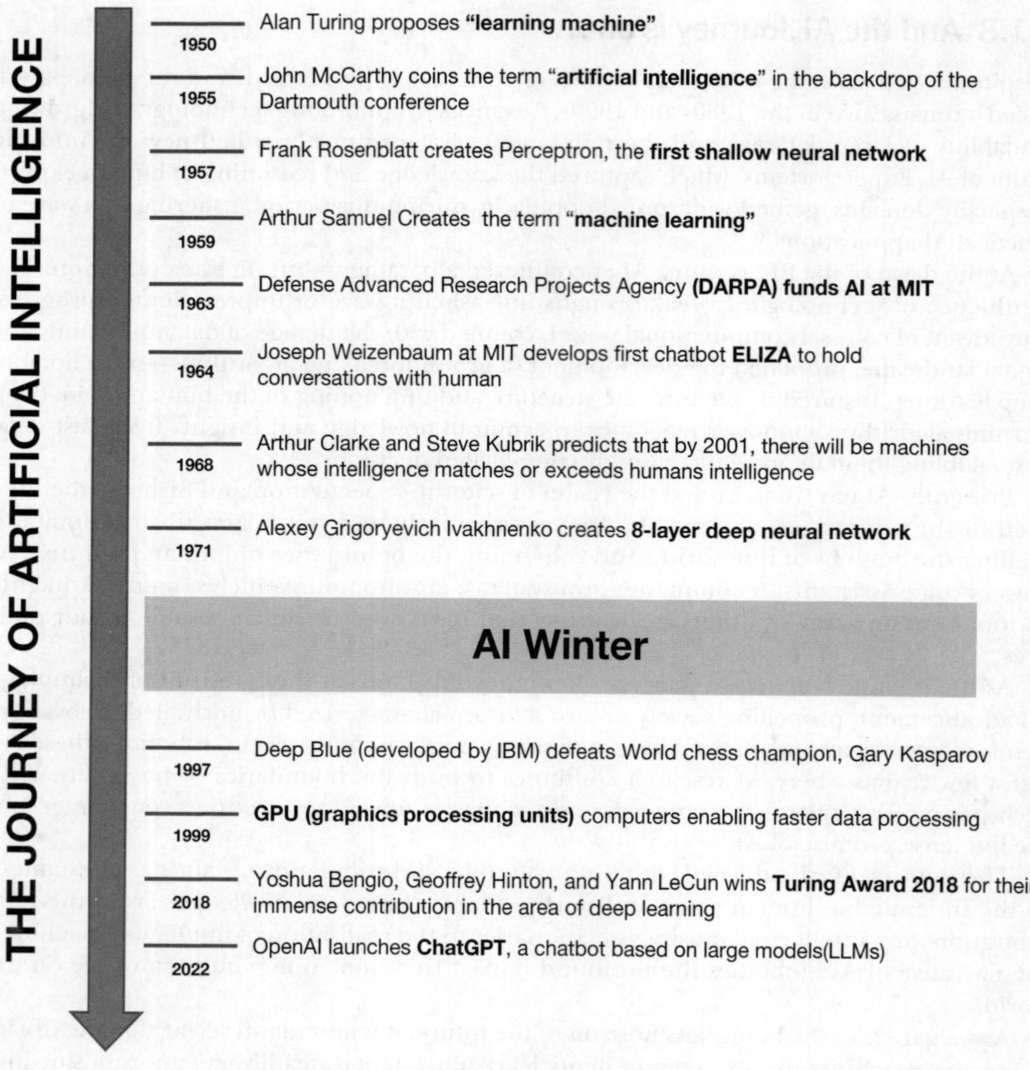

Figure 1.3 Journey of Artificial Intelligence Evolution

1.2 WHAT IS AI?

Think of Artificial Intelligence as a machine that can think and learn like a human but doesn't have a natural body. This machine can do intelligent things, make decisions and even help people with various tasks. The 'brain' of this machine is like a super-smart computer program. It's not made of natural brain tissue, but it's designed to think and learn from information, just like we do. This program uses lots of data and clever mathematics to make decisions. AI learns by looking at many examples and information, like how you learn from books, videos and experiences. Imagine you're teaching your robot friend to recognize different animals. You'd show it pictures of cats, dogs, and birds, and it would learn to tell them apart by studying those pictures.

More formally, AI is a multifaceted field that tries to emulate human-like intelligence in machines. A critical aspect of human intelligence is the ability to 'think'. The ability to think comes from multiple elements like the ability to understand, learn, do reasoning, communicate and act. Artificial intelligence is an area in computer science that tries to replicate human intelligence into machines and, in the process, builds intelligent machines. Authors Stuart Russell and Peter Norvig, in their famous textbook *Artificial Intelligence: A Modern Approach*, have called out four approaches that have been followed in AI:

- Thinking Humanly i.e., machines with minds
- Thinking Rationally i.e., machines perceive, reason and act
- Acting Humanly i.e., machines perform functions that require human intelligence
- Acting Rationally i.e., machines act autonomously to achieve the best possible outcome

Thus, AI tries to imbibe human-like cognitive abilities. At its core, AI aims to create intelligent systems that can think, learn, adapt and perform tasks typically requiring human intelligence. To comprehend the essence of AI, let's explore its essential facets.

1.2.1 Machine Learning: The Heart of AI

Imagine teaching a computer to recognize different types of fruit. AI does this by using a branch called machine learning. Just like you learn from examples, AI algorithms process vast amounts of data to improve their performance over time. With each piece of data, they get better at recognizing patterns, whether identifying spam emails or recognizing faces in photos.

1.2.2 Pattern Recognition: The Power to Distinguish

Pattern recognition is a superpower of AI. It helps machines find meaning in data. For instance, in healthcare, AI can spot patterns in medical images like X-rays or MRIs, aiding doctors in detecting diseases earlier and more accurately.

1.2.3 Natural Language Processing (NLP): Conversations with Computers

Think about talking to your computer like you speak to a friend. AI makes this possible through Natural Language Processing (NLP). It allows computers to understand and generate human language. Chatbots, virtual assistants like Siri or Alexa, and language translation apps rely on NLP.

1.2.4 Computer Vision: Seeing the World Through AI Eyes

Computer vision is a subset of AI that focuses on enabling machines to view and interpret the visual world. It involves teaching computers to understand and analyse images and videos. Computer vision has far-reaching applications, from helping robots navigate environments to facial recognition in smartphones and even assisting in medical diagnoses by interpreting medical images.

POINTS TO PONDER

- *AI-Generated Artwork Sold for $432,500:*
  In 2018, a piece of art titled 'Portrait of Edmond de Belamy' was created using a type of AI called Generative Adversarial Networks (GANs). This AI-generated artwork surprised the art world when it was sold at auction for $432,500, sparking discussions about the intersection of AI and creativity.
- *AI-Powered Robot Sophia Granted Citizenship:*
  In 2017, Saudi Arabia granted citizenship to Sophia, a humanoid robot developed by Hanson Robotics. This decision to give citizenship to a robot, raised questions about the legal status and rights of AI entities and ignited debates on AI ethics and citizenship.

Artificial Intelligence (AI) has become a part of the modern world, revolutionizing numerous industries and aspects of our daily lives. In all domains namely Banking, Insurance, Pharmaceuticals, Healthcare, Retail, etc., problems are solved using AI. We will look at a few such examples in the next section.

1.3 ARTIFICIAL GENERAL INTELLIGENCE (AGI)

Artificial General Intelligence (AGI), often referred to as 'strong AI' or 'full AI,' is a concept in artificial intelligence that represents a level of machine intelligence capable of understanding, learning, and performing any intellectual task that a human being can do. AGI aims to replicate human-like cognitive abilities, including reasoning, problem-solving, common-sense understanding, and adaptability across various domains. It is a highly advanced form of AI that can learn and apply knowledge in diverse situations, just as humans can.

1.3.1 Distinction from Narrow or Weak AI

Scope of Abilities

- **AGI:** AGI possesses a broad spectrum of cognitive capabilities and can excel in various tasks without specialized programming. It can switch tasks and adapt to new challenges, akin to human intelligence.
- **Narrow or Weak AI:** In contrast, narrow AI, also known as 'weak AI,' is designed for a specific task or a limited set of tasks. It excels in performing a particular function but lacks the flexibility and generalization capabilities of AGI.

Learning and Adaptation

- **AGI:** AGI systems can learn from experience and apply that knowledge to novel situations. They can generalize learning from one domain to another, showcasing adaptability and problem-solving skills.
- **Narrow or Weak AI:** Narrow AI systems are typically trained or programmed for a single task and do not possess the ability to generalize their learning beyond that specific task.

Human-like Understanding
- **AGI:** AGI strives to emulate human-like cognitive processes, understanding context, meaning, and nuances in language and behaviour. It aims to interact with humans that feel natural and intelligent.
- **Narrow or Weak AI:** Narrow AI systems lack proper understanding and operate based on predefined rules and patterns. They may excel in specific tasks, such as image recognition or language translation, but do not possess human-like comprehension.

Transferability
- **AGI:** AGI is transferable across various domains and can apply its knowledge to different situations, demonstrating versatility.
- **Narrow or Weak AI:** Narrow AI solutions are domain-specific and do not transfer their skills or knowledge to unrelated tasks or domains.

Self-improvement
- **AGI:** AGI systems have the potential for self-improvement, allowing them to enhance their abilities and acquire new skills independently.
- **Narrow or Weak AI:** Narrow AI systems require external modifications or reprogramming to expand their capabilities.

In summary, the concept of Artificial General Intelligence (AGI) represents the aspiration to create machines with human-like intelligence and adaptability, capable of performing diverse tasks. This stands in contrast to Narrow or Weak AI, which excels in specific, predefined tasks but lacks the broad cognitive abilities and versatility of AGI. Achieving AGI remains a significant challenge in artificial intelligence and represents a future goal that holds profound implications for technology, society and our understanding of intelligence.

1.4 INDUSTRY APPLICATIONS OF AI

AI continues to evolve and find new applications across diverse fields, reshaping industries and empowering us to tackle complex challenges in ways that were once unimaginable. Here are some critical applications of AI in various sectors.

1.4.1 Banking and Finance

There are multiple applications of AI in Banking and Finance industry, as depicted in Fig. 1.4. Few of the use cases have been highlighted below."

- **Fraud Detection:** AI helps banks identify unusual or suspicious transactions quickly. It can find certain patterns that humans might miss, helping to protect customers from fraud.
- **Customer Service:** AI-powered chatbots and virtual assistants can answer customer questions and provide support 24/7, making banking services more accessible.
- **Credit Scoring:** AI analyses a customer's financial history and behaviour to determine their creditworthiness, making it easier for banks to decide whether to approve loans or credit card applications.

Figure 1.4 Artificial Intelligence in Banking and Finance

- **Personalized Recommendations:** Banks can use AI to suggest personalized financial products and services based on a customer's spending habits and financial goals.
- **Risk Management:** AI helps banks assess and manage risks by analysing large amounts of data to make more informed decisions about investments and lending.
- **Automation:** Routine tasks, such as data entry and paperwork, can be automated with AI, reducing human error and saving time.
- **Security:** AI helps banks enhance cybersecurity by identifying potential threats and vulnerabilities in real-time, protecting sensitive customer data.
- **Predictive Analytics:** AI can predict market trends and customer behaviour, helping banks make better investment decisions and tailor their services.
- **Compliance and Regulation:** AI assists banks in ensuring they comply with financial regulations by monitoring transactions and identifying suspicious activities.

1.4.2 Insurance

Insurance industry has seen multiple applications of AI, as depicted in Fig. 1.5. Some salient use cases are presented below.

- **Claims Processing:** AI helps insurance companies quickly process claims by analysing documents and photos to assess damage or injuries, making the process faster and more accurate.
- **Risk Assessment:** AI analyses data to determine the level of risk associated with insuring a person or property, which helps insurance companies set appropriate premiums.
- **Customer Service:** AI-powered chatbots and virtual assistants assist customers with questions, policy information and claims, providing 24/7 support.
- **Fraud Detection:** AI identifies suspicious behaviour or false claims, helping insurance companies prevent and investigate fraud.
- **Underwriting:** AI evaluates applicant's information to determine if they qualify for insurance and at what cost, making the underwriting process more efficient.
- **Predictive Analytics:** AI predicts future trends and risks, helping insurance companies make informed decisions about coverage and pricing.

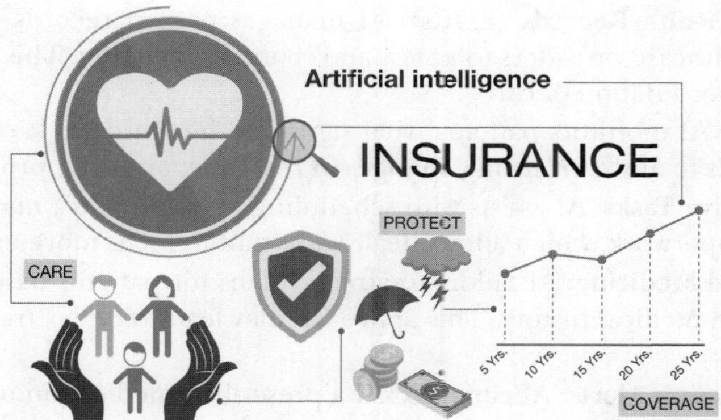

Figure 1.5 Artificial Intelligence in Insurance

- **Personalized Policies:** AI allows insurers to offer customized insurance plans based on an individual's unique needs and circumstances.

1.4.3 Healthcare

Healthcare is another sensitive domain which finds a number of applications of AI, a sample depiction shown in Fig. 1.6. A summary of some key uses of AI in Healthcare is presented below.

- **Diagnosis and Imaging:** AI helps doctors by analysing medical images like X-rays, CT scans, and MRIs. It can highlight potential health issues, making it easier for doctors to spot problems early.
- **Treatment Recommendations:** AI suggests treatment options based on a patient's medical history and current condition. It can guide the most effective therapies and medications.
- **Drug Discovery:** AI speeds up finding new medicines and treatments. It analyses a significant volume of data to identify potential drug candidates, saving time and resources.

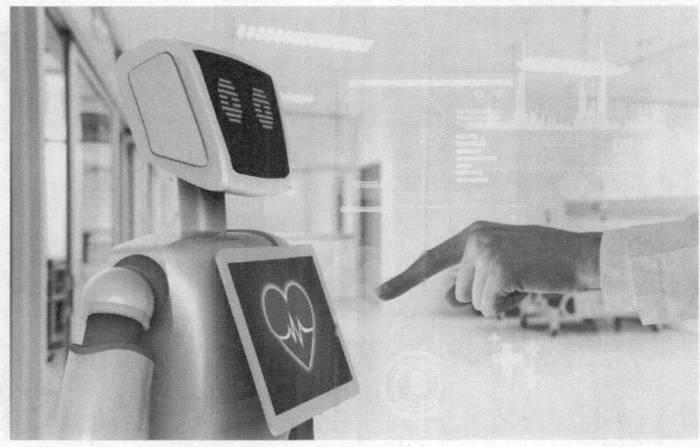

Figure 1.6 Artificial Intelligence in Healthcare

- **Electronic Health Records (EHRs):** AI manages patient records digitally, making it easy for healthcare providers to access and share information. This reduces errors and improves coordination of care.
- **Monitoring:** AI monitors patients' vital signs and health data. It can alert healthcare professionals to any concerning changes in real-time, ensuring prompt intervention.
- **Administrative Tasks:** AI assists with scheduling appointments, managing billing, and handling paperwork, which allows healthcare staff to focus more on patient care.
- **Personalized Medicine:** AI tailors treatment plans for patients based on their genetic makeup and medical history. This approach may lead to better treatments with fewer side effects.
- **Drug Interaction Alerts:** AI can check if a prescribed medication might interact negatively with other drugs a patient is taking, preventing potential harm.
- **Healthcare Chatbots:** AI-powered chatbots can answer patients' questions, provide information about symptoms, and offer guidance on when to seek medical attention. They are available 24/7.
- **Disease Prediction:** AI can analyse health data and predict the likelihood of diseases or certain health conditions in individuals, allowing for early intervention and prevention.

1.4.4 Retail and E-commerce

Number of applications of AI is also found in the Retail and E-commerce domain, as depicted in Fig. 1.7. Few of the salient use cases have been highlighted below.

- **Product Recommendations:** AI algorithms analyse a customer's purchase history, browsing behaviour and demographic information to suggest relevant products. This personalization increases the likelihood of making a sale and enhances the customer's shopping experience.

Figure 1.7 Artificial Intelligence in Retail and E-commerce

- **Inventory Management:** AI helps retailers optimize their inventory by predicting demand patterns, identifying slow-moving items and automatically reordering products when stock levels are low. This reduces overstocking and understocking issues, ultimately saving costs.
- **Customer Service:** AI-powered chatbots and virtual assistants are available 24/7 to answer customer inquiries, provide product information, and assist with basic problem-solving. They can handle routine queries, freeing human customer service agents for more complex issues.
- **Price Optimization:** AI algorithms analyse various factors such as competitor pricing, demand fluctuations and historical sales data to adjust product prices in real-time. This dynamic pricing strategy maximizes profits and competitiveness.
- **Visual Search:** AI enables customers to search for products using images or photos instead of text queries. Optical search technology can identify and match items based on their observable characteristics, making it easier for customers to find the products they want.
- **Fraud Detection:** AI algorithms continuously monitor transactions and customer behaviours to identify unusual or suspicious activities, such as fraudulent payment attempts or account hijacking. This proactive approach helps prevent financial losses for both customers and businesses.
- **Supply Chain Optimization:** AI optimizes supply chain operations by predicting demand, improving logistics and transportation routes, and reducing lead times. This results in cost savings and faster order fulfilment.
- **Customer Insights:** AI analyses vast amounts of customer data to gain insights into preferences, buying patterns and trends. Retailers can use these insights to tailor marketing campaigns, product assortments, and store layouts to meet customer needs better.
- **Chatbots for Sales:** Beyond customer service, AI-powered chatbots can assist with sales by guiding customers through product selection, providing product recommendations, and even processing orders directly within the chat interface.
- **Personalized Marketing:** AI enables retailers to create personalized marketing campaigns through targeted advertisements, email recommendations and content curation, increasing the likelihood of conversion and customer loyalty.

1.4.5 Manufacturing

As depicted in Fig. 1.8, AI finds it's applications in the Manufacturing industry too in the form of robotic automation and other use cases. Few of the salient applications have been highlighted below.

- **Quality Control:** AI checks products as they are made, looking for defects like cracks or errors. This ensures that items are made to a high standard and reduces waste.
- **Predictive Maintenance:** AI predicts when machines might break down and need repairs. This prevents costly unexpected shutdowns and keeps production running smoothly.
- **Production Optimization:** AI helps factories plan and manage production schedules. It can adjust production rates based on demand, reducing excess inventory.

Figure 1.8 Artificial Intelligence in Manufacturing

- **Supply Chain Management:** AI tracks the movement of materials and parts in real-time, ensuring they arrive at the right place at the right time. This keeps production lines running efficiently.
- **Robots and Automation:** AI controls robots that assemble, weld, or move factory materials. These robots work tirelessly and precisely, speeding up production.
- **Safety:** AI monitors the safety of factory environments. It can detect dangerous conditions and shut down machines if there's a risk to workers.
- **Inventory Management:** AI manages how much raw material is needed and when to order it. This helps factories save money by reducing excess inventory.
- **Energy Efficiency:** AI helps factories use energy more efficiently, reducing costs and environmental impact.
- **Customization:** AI can adapt production lines to make customized products without slowing down manufacturing.
- **Data Analysis:** AI looks at lots of data to find ways to improve processes and make products better and cheaper.

1.4.6 Entertainment

Diverse applications of AI are found in the Entertainment industry too in the form content generation and other use cases, as depicted in Fig. 1.9. Highlights of some of the key applications are presented below.

- **Content Recommendation:** AI suggests movies, shows, music, and games you might like based on what you've enjoyed before. It's like having a friend who knows your taste.
- **Content Creation:** AI can generate music, art and even write stories. It's used to create special effects and animations in movies, making them look more impressive.
- **Personalized Experiences:** AI in video games adapts to your skills and preferences. It makes the game challenging but not too hard, so you have the most fun.

Figure 1.9 Artificial Intelligence in Entertainment

- **Streaming Quality:** AI ensures that streaming services like Netflix and YouTube run smoothly without buffering, so that you can enjoy your favourite shows without interruption.
- **Voice and Face Recognition:** AI powers virtual assistants like Siri and Alexa. They listen to your voice commands and can even recognize your face for security.
- **Interactive Chatbots:** Some websites and social media use AI-powered chatbots to have fun conversations or help you find information.
- **Editing and Post-Production:** AI helps in editing videos, adding special effects, and improving the overall quality of content.
- **Predictive Analytics:** Entertainment companies use AI to predict what movies or songs might become popular, helping them decide what to produce next.
- **Enhanced Virtual Reality (VR):** AI improves the realism and immersion in VR experiences, making games and simulations more exciting.
- **Copyright Protection:** AI helps identify and prevent the illegal distribution of movies, music and other entertainment content.

1.4.7 Agriculture

Agriculture also finds multiple applications of AI in the form of precision farming, crop monitoring and so on. A representative depiction has been presented in Fig. 1.10. Few of the use cases related to application of AI in Agriculture have been highlighted below.

- **Crop Monitoring:** AI uses drones and sensors to check on crops. It looks for signs of pests, disease, or if they need more water or nutrients. This helps farmers take action early to protect their plants.
- **Precision Farming:** AI helps farmers decide precisely where and when to plant seeds and apply fertilizers. This saves money and makes farming more efficient.

Figure 1.10 Artificial Intelligence in Agriculture

- **Weather Predictions:** AI analyses weather data to provide accurate forecasts. Farmers can plan their work better and protect their crops from extreme weather.

- **Harvesting Robots:** AI-powered robots can pick fruits and vegetables without damaging them. This makes harvesting faster and more cost-effective.

- **Livestock Management:** AI monitors the health of animals, tracking things like body temperature and weight. This helps farmers keep their animals healthy.

- **Soil Analysis:** AI checks the soil quality and recommends the right type and amount of fertilizers. This helps ensure healthy plant growth.

- **Crop Sorting and Grading:** AI sorts and grades harvested crops based on their size, quality, and ripeness, making it easier to sell them at the right price.

- **Pest Control:** AI identifies harmful insects and weeds. It can suggest eco-friendly ways to control them, reducing the need for toxic pesticides.

- **Supply Chain Management:** AI tracks the movement of crops and animals from the farm to the market. This ensures fresh and safe food reaches consumers.

- **Farm Management Software:** AI-powered apps help farmers with planning, record-keeping, and decision-making, making farming more organized and profitable.

1.4.8 Education

Education domain has seen a paradigm shift with the application of AI. A sample depiction is presented in Fig. 1.11. Few of the use cases have been highlighted below.

- **Personalized Learning:** AI customizes lessons for each student. It figures out what you're good at and what you need help with, so you learn better.
- **Tutoring:** AI-powered programs can explain tricky concepts and answer questions, just like a tutor. They're available 24/7, so you can get help whenever needed.
- **Grading and Feedback:** AI can grade assignments and tests quickly, giving instant feedback. Teachers have more time to help you improve.
- **Curriculum Planning:** AI helps teachers choose the best materials and methods for teaching. It keeps lessons interesting and compelling.
- **Language Translation:** AI can translate languages, making it easier for students worldwide to access educational content.
- **Accessibility:** AI can assist students with disabilities by providing text-to-speech, speech-to-text, or other tools to make learning more accessible.
- **Data Analysis:** AI looks at lots of data to spot trends. It helps schools see what's working and what needs improvement.
- **Virtual Classrooms:** AI powers virtual classrooms where students and teachers can interact online. This is especially useful for distance learning.
- **Study Aids:** AI can create study guides and practice tests tailored to what you need to learn.
- **Time Management:** AI helps students and teachers organize schedules and deadlines, ensuring everyone stays on track.

Figure 1.11 Artificial Intelligence in Education

1.5 CHALLENGES IN AI

Artificial Intelligence (AI) holds immense promise, but it also presents diverse challenges that need careful consideration. As AI technologies continue to advance and integrate into various aspects of our lives, addressing these challenges is essential. From data privacy concerns to ethical dilemmas and the need for transparency, this list provides a detailed exploration of the critical challenges that AI faces in its quest to deliver benefits while mitigating risks.

1.5.1 Data Privacy

- AI systems rely on vast amounts of data, often personal and sensitive information. Safeguarding this data from breaches and ensuring individuals' privacy rights are protected is a complex task.
- Ensuring data is anonymized, obtaining informed consent and establishing robust encryption methods are critical. Striking a balance between data access for AI and user privacy remains challenging.

1.5.2 Bias and Fairness

- AI algorithms can perpetuate biases present in historical data, leading to unfair outcomes and reinforcing discrimination.
- Identifying and mitigating biases in AI models, setting fairness standards, and ensuring diverse representation in AI development are ongoing challenges.

Figure 1.12 Ethical considerations in AI — Balancing progress and responsibility

1.5.3 Transparency and Explainability

- Many AI models, like deep neural networks, lack transparency, making it difficult to comprehend how they reach conclusions.
- Developing methods to interpret AI model decisions, creating explainable AI techniques and fostering transparency in AI development are active areas of research and regulation.

1.5.4 Ethical Concerns

- AI raises ethical dilemmas in various applications, such as the use of facial recognition in surveillance, or AI in autonomous weaponry.
- Establishing ethical guidelines for AI use, addressing concerns about bias and discrimination, and determining responsible AI practices in various domains, are complex ethical challenges.

1.5.5 Job Displacement

- Automation driven by AI can lead to concerns about job losses in industries like manufacturing and customer service.
- Preparing the workforce for AI-driven changes, offering retraining programs and addressing potential socioeconomic disparities are critical considerations.

1.5.6 Regulation and Standards

- AI evolves rapidly, making it challenging for regulations and standards to keep pace.
- Developing comprehensive AI regulations, international cooperation on AI governance, and fostering industry-wide standards are ongoing challenges.

1.5.7 Security Risks

- AI systems can be vulnerable to attacks and adversarial manipulation.
- Ensuring robust cybersecurity measures, protecting against AI-generated threats and detecting vulnerabilities in AI systems are ongoing security challenges.

1.5.8 Resource Intensiveness

- Training complex AI models consumes significant computational resources, leading to environmental concerns and accessibility issues.
- Exploring energy-efficient AI algorithms, promoting responsible AI infrastructure and improving resource accessibility are essential steps.

1.5.9 Lack of Skilled Workforce

- There is a shortage of professionals with AI expertise.
- Expanding educational opportunities, offering AI training programs and fostering AI research and development talent are vital actions.

1.5.10 Interoperability

- Integrating AI systems with existing technologies and data sources can be complex.
- Developing standardized interfaces, ensuring data compatibility, and promoting interoperability across AI applications and platforms are critical.

1.5.11 Accountability

- **Challenge:** Determining responsibility in cases of AI errors or harm is complex, as multiple parties may be involved.
- **Detailed Concerns:** Establishing clear accountability frameworks, liability rules, and ethical guidelines for AI developers, users and regulators are ongoing efforts.

Addressing these detailed challenges requires ongoing collaboration among governments, industry stakeholders, researchers, and the public to shape AI technologies and policies responsibly.

1.6 CONCLUSION

In conclusion, the journey into Artificial Intelligence is nothing short of an exciting journey, marked by the convergence of human ingenuity, technological prowess, and boundless curiosity. As we've traversed the landscape of AI in this introductory chapter, we've embarked on a voyage of discovery, uncovering the immense potential and captivating challenges that this field presents. At its core, AI represents a quest to replicate and enhance human intelligence, offering a glimpse into what it means to be intelligent and adaptable.

We began by unravelling the foundational principles of AI, exploring its fundamental concepts, and understanding the mechanics of machine learning and neural networks. Through this foundational knowledge, we gained insights into how machines can learn from data, recognize patterns and make predictions—a process that mirrors the way our minds work. It's a testament to our capacity to teach machines, not just what to think, but how to think - a transformative power with far-reaching implications.

Moreover, our journey has taken us beyond the theoretical underpinnings of AI into the realm of practical applications reshaping the world around us. We've witnessed AI's influence across various industries, from healthcare and finance to entertainment and agriculture. It's clear that AI isn't confined to the realm of science fiction but is increasingly becoming a cornerstone of our daily lives. It streamlines processes, enhances decision-making, and augments human capabilities, offering a glimpse into the future of what's possible when human creativity and AI collaboration intertwine.

Yet, as with any transformative force, AI brings with it a set of intricate challenges and ethical considerations. Issues surrounding data privacy, bias, transparency and accountability loom large on the horizon, demanding our collective attention and thoughtful solutions. We must navigate these complexities with care, ensuring that AI continues to serve humanity's best interests while mitigating potential risks.

As we look ahead to the following chapters, we are poised to delve even deeper into the heart of AI. We'll journey through its historical origins, trace its evolution, and witness the cutting-edge breakthroughs that shape our future. The chapters to come will be a testament to the unceasing spirit of exploration and innovation that drives the field of AI. Each revelation, each breakthrough, takes us one step closer to unlocking the full potential of human intelligence and harnessing AI as a powerful tool for solving complex problems, creating new possibilities, and reshaping our world.

So, let us wholeheartedly embrace this remarkable journey that lies ahead—a journey into the captivating world of Artificial Intelligence. It's a journey where every horizon expands with every discovery, every innovation, and every collaboration between human genius and the computational brilliance of AI. As we venture forth, we carry with us the torch of curiosity and the aspirations of a future that is made brighter, smarter and more wondrous by the symbiotic relationship between human intellect and AI's transformative potential.

1.7 SUMMARY

- The birth of AI as a formal area can be traced back to 1956 - the Dartmouth Conference. It was in this conference that the term 'artificial intelligence' was first uttered.

- The 1940s and 1950s were formative years for AI research, characterized by foundational work in computer science.

- In 1943, McCulloch and Pitts collaborated to create a mathematical model of a simplified artificial neuron, often referred to as the 'McCulloch-Pitts neuron'.

- Alexey Grigoryevich Ivakhnenko, considered to be the Father of Deep Learning developed the first multi-layer perceptron in 1965, which forms the basis of deep learning.

- The first AI winter occurred in the early 1970s. Funding for AI research decreased, and many AI projects were discontinued.

- The 1980s represented a recovery period following the AI winter of the 1970s. The 1980s saw an increase in the development and adoption of expert systems across various industries.

- The 1990s marked a resurgence of interest in machine learning, a subfield of AI focused on creating algorithms that could learn from data. Reinforcement learning, a type of machine learning, also gained prominence.

- In 1997, IBM's Deep Blue supercomputer famously defeated world chess champion Garry Kasparov.

- The first decade of the 21st century have marked a pivotal period of artificial intelligence (AI) research, with several vital incidents reshaping the field and propelling it into the mainstream.

- The second decade of the 21st century have witnessed an unprecedented surge in AI research and development.

- AI is a multifaceted field that tries to emulate human-like intelligence in machines.

- Artificial General Intelligence (AGI), often referred to as 'strong AI' or 'full AI,' is a concept in artificial intelligence that represents a level of machine intelligence capable of understanding, learning, and performing any intellectual task that a human being can do.

- AI is successfully used in multiple industries reshaping the industries and solving complex problems.

- Artificial Intelligence (AI) holds immense promise, but it also presents diverse challenges ranging from data privacy concerns to ethical dilemmas, that need careful consideration.

1.8 PRACTICE EXERCISES

1.8.1 Subjective Questions (10 Marks)

1. Define artificial intelligence (AI) and elaborate on its significance in contemporary society. Discuss how AI differs from traditional computer programming, and provide examples of real-world AI applications across different industries.

2. Trace the historical development of AI from its inception to the present day. Highlight key milestones, breakthroughs, and the evolution of AI concepts, and explain how these have shaped the field.

3. Explain the basic principles of machine learning and deep learning within the context of AI. Provide examples of how machine learning algorithms work, their applications, and the importance of data in training AI models.

4. Discuss the concept of artificial general intelligence (AGI) and its distinction from narrow or weak AI. Explore the challenges and ethical considerations associated with AGI development.

5. Explore both the opportunities and challenges posed by integrating AI technologies into various industries.

6. Discuss the ethical implications of AI, including issues related to bias, privacy and accountability. Provide examples of AI ethics frameworks and initiatives aimed at addressing these concerns.

7. Examine the role of AI in healthcare, including its applications in diagnosis, treatment, and medical research. Assess the potential benefits and challenges of AI adoption in the healthcare sector.

8. What are the potential societal impacts of widespread AI adoption? Discuss the effects of AI on labour markets, economic inequality and the redistribution of job roles.

9. Explore the concept of AI ethics and its importance in guiding responsible AI development and deployment. Provide examples of AI ethics guidelines and principles established by organizations and governments.

10. Compare and contrast AI, machine learning and deep learning. Provide examples of tasks that each type of learning is suited for.

1.8.2 Subjective Questions (5 Marks)

1. Explain the significance of the Dartmouth Workshop in the history of AI.

2. Describe the impact of the 'AI winter' on the progression of artificial intelligence and its resurgence in later years.

3. Analyse the historical evolution of natural language processing (NLP) in AI research and its significance in contemporary AI applications.

4. Discuss the significant advancements in machine learning during the 2010s and their influence on the AI landscape.

5. Describe the development of early neural networks, such as the perceptron, and their contributions to AI.

6. Explain the concept of 'AI ethics' and its growing importance in the context of AI research and deployment. Provide examples of ethical concerns related to AI.

7. Discuss the impact of AI on various industries, such as healthcare, finance, and transportation, and highlight specific use cases within these sectors.

8. Investigate the role of AI in addressing grand global challenges, such as pandemics, climate change and food security. How can AI be harnessed to provide innovative solutions to these complex problems?

9. In the context of AI ethics, what are the dilemmas and trade-offs involved in balancing issues like privacy, transparency, fairness, and security in AI systems?

10. Discuss the ethical and legal challenges surrounding the ownership and governance of AI-generated intellectual property, such as art, music, or literature created by AI systems.

1.8.3 Short Answer-type Questions (2 Marks)

1. When did the concept of artificial intelligence (AI) first emerge?

2. Who is considered the 'father of artificial intelligence'?

3. What is the Turing test, and why is it significant in AI?

4. Name one of the earliest AI programs created in the 1950s.

5. What is machine learning and how does it relate to AI?

6. What is the significance of neural networks in modern AI?

7. What role does data play in training AI models?

8. Who developed the first AI chess-playing computer program, and when?

9. How does natural language processing (NLP) contribute to AI applications?

10. What are the primary challenges in achieving artificial general intelligence (AGI)?

11. What are the potential societal impacts of widespread AI adoption?

12. What are the differences between narrow or weak AI and strong AI?

13. How would you define artificial intelligence (AI) in simple terms?

14. Can you provide an example of a task that AI can perform?

15. What distinguishes AI from traditional computer programming?

16. What are the critical components of an AI system?

17. What are some ethical considerations in the use of AI technologies?

18. What are the potential benefits of integrating AI into various industries?

19. What are the challenges in achieving human-level AI intelligence?

20. What are some common misconceptions about AI?

1.8.4 Multiple-choice Questions (1 Mark)

1. What event is often considered the birth of artificial intelligence (AI)?
   (a) The invention of the first computer
   (b) The Dartmouth Workshop in 1956
   (c) Alan Turing's Turing Test
   (d) The development of the first neural network

2. Who coined the term 'artificial intelligence' during the Dartmouth Workshop in 1956?
   (a) John McCarthy
   (b) Alan Turing
   (c) Marvin Minsky
   (d) Frank Rosenblatt

3. Which early AI model aimed to mimic the functioning of a biological neuron?
   (a) The Turing Machine
   (b) The Logic Theorist
   (c) The McCulloch-Pitts neuron
   (d) The Group Method of Data Handling (GMDH)

4. What is the primary function of a perceptron in AI?
   (a) Image recognition
   (b) Natural language processing
   (c) Pattern recognition
   (d) Speech synthesis

5. During which period did the 'AI winter' occur?
    (a) 1940s and 1950s
    (b) 1960s and 1970s
    (c) 1980s and 1990s
    (d) 2000s and 2010s

6. What concept did Frank Rosenblatt's perceptron contribute to the field of AI?
    (a) Deep learning
    (b) Reinforcement learning
    (c) Symbolic logic
    (d) Machine learning

7. Which AI algorithm defeated the world Go champion in 2016, demonstrating AI's strategic capabilities?
    (a) Deep Blue
    (b) Watson
    (c) AlphaGo
    (d) GPT-3

8. What is the primary function of a convolutional neural network (CNN) in AI?
    (a) Natural language processing
    (b) Image recognition
    (c) Stock market prediction
    (d) Weather forecasting

9. What important ethical concern has emerged with the widespread use of AI?
    (a) Data privacy
    (b) AI winter
    (c) Moore's Law
    (d) The Turing Test

10. What is the primary purpose of the Turing Test in AI?
    (a) To evaluate a machine's ability to think like a human
    (b) To assess a machine's computational speed
    (c) To test a machine's memory capacity
    (d) To measure a machine's creativity

11. In AI, what does 'NLP' stand for?
    (a) Neurological Linguistic Programming
    (b) Natural Language Programming
    (c) Neural Learning Protocol
    (d) Natural Language Processing

12. Which AI application uses algorithms to analyse and interpret human emotions from text, voice, or facial expressions?
    (a) Sentiment analysis
    (b) Speech recognition
    (c) Object detection
    (d) Reinforcement learning

13. What is the primary goal of AI in autonomous vehicles (self-driving cars)?
    (a) To increase fuel efficiency
    (b) To reduce manufacturing costs
    (c) To enhance passenger comfort
    (d) To enable safe and efficient navigation

14. What is a typical application of AI in the field of healthcare?
    (a) Sentiment analysis
    (b) Disease diagnosis
    (c) Weather forecasting
    (d) Video game development

15. Which AI technology allows machines to recognize and understand human speech?
    (a) Computer vision
    (b) Natural language processing
    (c) Reinforcement learning
    (d) Genetic algorithms

16. What is the primary goal of AI-driven recommendation systems, such as those used by Netflix or Amazon?
    (a) To optimize server performance
    (b) To increase website traffic
    (c) To personalize content and product recommendations
    (d) To reduce data storage costs

1.9 ANSWER KEYS

1.9.1 Multiple-choice Questions

1. (b)	2. (a)	3. (c)	4. (c)	5. (b)	6. (d)	7. (c)	8. (b)
9. (a)	10. (a)	11. (d)	12. (a)	13. (d)	14. (b)	15. (b)	16. (c)

CHAPTER 2

AI Subfields

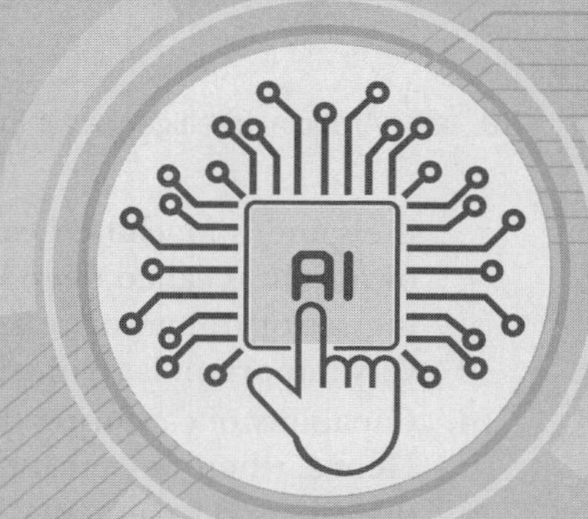

2.1 INTRODUCTION

In Chapter 1, we have been introduced to the basic concepts of Artificial Intelligence. In this chapter, we will concentrate on different constituents of AI and a brief outline about each one of them. The grand aim or the ultimate vision of AI has been to simulate human thinking. Before moving forward, it is also essential to differentiate between intelligent thinking and intelligent action. The below table gives a very high-level difference between the two.

Intelligent Thinking	Intelligent Action
Intelligent thinking, also known as cognitive intelligence, refers to the ability of AI systems to simulate human thought processes, reasoning, and problem-solving. **Examples:** Understanding natural language, recognizing patterns, making inferences, and solving complex problems	Intelligent action, also known as behavioural intelligence, pertains to the ability of AI systems to interact with and manipulate the physical world. **Examples:** Decision-making, planning and execution of actions

Hence, a more precise goal of AI is to develop systems which can exhibit human-level or superhuman-level performance in a wide range of cognitive and physical tasks. A more tangible goal can also be to build an intelligent robot.

What are some of the capabilities that such a system will need?

1. Capability to make sense of visual cues (for example, reading the traffic signs in case of a self-driving car). The subfield dedicated to the above is called **computer vision**.
2. Capability to understand natural language and respond in natural language (one of the tasks that is routinely used here is creating a summary of a given text). Formally such a task is called text summarization. The subfield of AI, that deals with the same is called **natural language processing.**
3. Capability of self-learning (When we want to learn anything new, we often look at a worked-out example and derive some rules in our mind). These examples are often called training data as per computational parlance. The subfield that concentrates on

the same is **Machine Learning**. A recent advancement, which deserves mention in its own right is called deep learning. This particular discipline is based on a very fascinating set of frameworks named Deep Neural Networks. One other technique which is very important is reinforcement learning.

4. Capability of capturing the knowledge in some way may be in the forms of some rules. Hence, when we consult a doctor, he has some concrete knowledge about symptoms, diseases and treatments. These construct a knowledge base. Logical Inferences may be made using the knowledge base. Some of the logical formalisms are **predicate logic, propositional logic, first-order logic,** etc. There is another very interesting dimension here. It turns out that in the real world, some information is not measurable but conveys sense, like this is very hot today or this vegetable is quite cheap so on and so Fuzzy Logic, Rough Set deals with the same.

5. Capability of acting on the basis of perceived inputs. Essentially, it is not enough to process a traffic signal showing the speed limit, the break needs to be pressed to slow down. The subfield that deals with the same is called robotics.

At a high level, these capabilities translate to the important subfields of AI. These subfields often intersect and collaborate, and advancements in one subfield often benefit others. AI is a rapidly evolving field, and new subfields and interdisciplinary areas continue to emerge as research progresses.

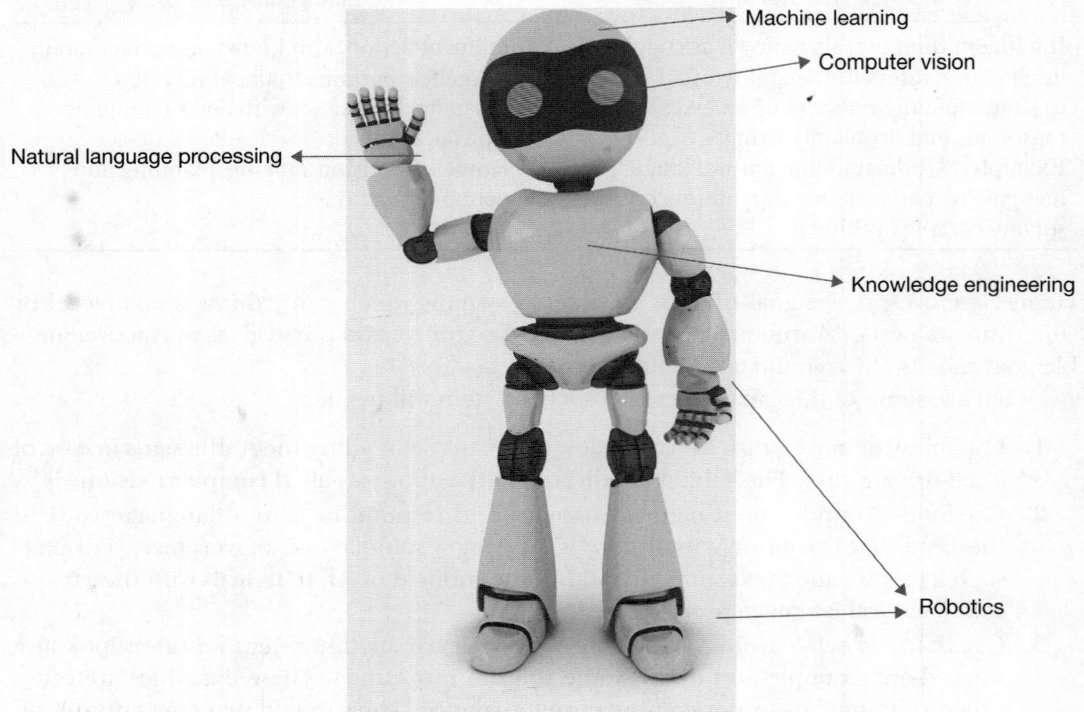

Figure 2.1 Different sub-fields with respect to an artificial man

A tabular summary of the subfields is enclosed below.

Subfield	Description	Applications
Computer vision	Extracts meaning from images and videos.	Object detection, facial recognition, scene understanding
Natural language processing	Deals with the interaction between computers and human languages.	Text classification, machine translation, sentiment analysis
Machine learning	Develops algorithms that can learn from data. Deep learning is a subfield of machine learning.	Image recognition, natural language processing, fraud detection
Knowledge engineering	Involves representation of knowledge and application of logic for inference.	Expert systems
Robotics	Develops robots that can interact with the physical world.	Navigation, manipulation, object recognition

There are many associated issues like explainability of the decisions, fairness and ethical considerations of the AI systems, which will be covered in subsequent chapters.

DO YOU KNOW ?

INDIAai, the National AI Portal of India, is a government-run website launched in May 2020 to provide a central hub for all things related to AI in India. It offers a variety of resources, including news and articles, information on key AI players, insights into the global and Indian AI landscape, resources for students, entrepreneurs, professionals and executives, and a database of AI initiatives and startups in India. INDIAai also plays a key role in promoting AI research and innovation in India by supporting cutting-edge research and applied activities on AI-related priorities.

2.2 COMPUTER VISION

Computer vision is a field of artificial intelligence that gives computers the ability to see, interpret and make sense of the world around them. However, fascinating this sounds it was a herculean task to capture, store, process and finally make sense of the images. Thanks to phenomenal development in storage, processors as well as algorithms, it is a rapidly growing field with a wide range of applications which is discussed hereafter.

2.2.1 Object Detection

When humans look at a photograph or watch a video, they have the innate ability to readily spot people, objects, scenes and visual details. The goal of object recognition is to teach a computer to do what comes naturally to humans: to gain a level of understanding of what an image contains. For example, a computer vision system could be used to detect cars in a traffic scene or people in a crowd.

Figure 2.2 Object Recognition

Typically, object recognition consists of two main steps. The first step is called object localization, where a rectangle is put around the objects as shown in Figure 2.2. Such a rectangle is known as the bounding box. The second step is object classification. Bounding Boxes for object recognition in an office environment is also shown in the above figure.

Hence, step 1, concentrates on ascertaining whether there is an object in the scene, if yes what type of object it is. Following are some of the examples where object recognition is used across the globe:

- The Mumbai Traffic Police is using object recognition to detect vehicles and pedestrians in traffic. This is done by using cameras to scan the roads and identify objects that are not supposed to be there. If a vehicle or pedestrian is detected, the traffic police may be alerted for taking action to prevent an accident. They have been one of the early adaptors of technologies by installing a large number of CCTV Cameras.
- AI-driven surveillance robots are currently under development by nations such as South Korea and Israel, with the objective of deploying them to oversee border fence security. Here object recognition is being applied for human detection.

- Drishti, a real-time object recognition application developed by researchers and scientists at the Indian Institute of Technology Bombay, serves as a valuable tool for assisting individuals with visual impairments in navigating their surroundings. This innovative application utilizes the smartphone's camera to identify objects within the user's environment, subsequently delivering spoken notifications to describe these objects. For instance, the app can inform users about their proximity to a road crossing, the presence of individuals in their path, or the existence of nearby staircases. Although Drishti remains in the development phase, it has already made a positive impact, with over 10,000 individuals in India having utilized it. Users have lauded the application for its accuracy and user-friendly interface.

2.2.2 Facial Recognition

This is the ability to identify a person's face from an image or video. Facial recognition systems are often used for security applications, such as access control and fraud detection. This is easier said than done, as a person may appear in front of a camera in various attires (with a cap, a sunglass, mask and what not), different backgrounds and varying luminosity. Below are two important implementations of facial recognition.

The Aadhaar system, managed by the Unique Identification Authority of India (UIDAI), is a biometric identification program that includes facial recognition as one of its authentication methods. Aadhaar cards, which contain unique identification numbers, are linked to individuals' biometric data, including facial images, for identity verification in various government and private sector services.

DigiYatra is an initiative by the Ministry of Civil Aviation in India aimed at creating a seamless travel experience for air passengers. It includes a facial recognition-based system for biometric authentication at airports. Passengers who voluntarily enrol in the program can go through various checkpoints, such as check-in, security and boarding gates, using facial recognition instead of traditional identification documents.

Hyderabad Police has more than 7,00,000 Cameras installed in the city. They are using these cameras, transmitters and finally a data centre to identify known criminals in the streets of the city. The system is going through a lot more developments as of now. Of course, Aadhaar is being used in authentication using face recognition.

Before we move on to other important areas of computer vision, let us clearly differentiate two terms namely face detection and face recognition. The following table summarizes the key differences.

Characteristic	Face detection	Face recognition
Purpose	To identify the presence of a human face in an image or video.	To identify the identity of a person based on their face.
Difficulty	More fundamental and general.	More advanced and specific.
Example	Identifying the faces of your friends in a photo.	Identifying the names of your friends in a photo.

2.2.3 Scene Understanding

This is the ability to understand the context of an image or video. For example, a computer vision system could be used to understand the layout of a room or the activities that are taking place in a scene.

Scene understanding is a more complex task, then the above two. Essentially, if a picture is shown, the textual description of the image will be required for understanding the scene. For example, there may be a scene where Virat Kohli is hitting Stuart Broad for a six, over long on. As you see this will involve detecting the persons, identifying them, detecting other objects, detecting and identifying the location and the activity.

There are a lot of applications of scene understanding. However, the most prominent of them are self-driving cars and AR/VR.

- **Self-driving cars:** Self-driving cars need to have the capability to understand the scene around them in order to make safe driving decisions.
- **Virtual reality and augmented reality:** Virtual reality and augmented reality applications need to be able to understand the scene around the user in order to create a realistic and immersive experience.

2.2.4 Medical Imaging

Computer vision is used extensively in medical imaging to analyse images of the human body. For example, systems/algorithms based on computer vision can be used to detect cancer cells in images of tissue or to diagnose diseases from X-rays. There are several imaging techniques like X-Rays, Computed Tomography (CT), Magnetic resonance imaging (MRI), Ultrasound, Positron emission tomography (PET), etc., which are routinely performed to help with diagnosis and effective treatment. AI techniques can be applied to assist doctors, this can also be helpful in a country like India where medical practitioners are in short supply. The initial interpretation can be done by an AI-based medical diagnosis system and then only priority cases can be routed to medical professionals.

For interested readers, there is a dedicated society the Medical Image Computing and Computer Assisted Intervention Society (MICCAI - http://www.miccai.org/). It also organizes a very reputed conference with the same name.

The pandemic has created a lot of interest in computer vision-based medical imaging techniques. Some of the benefits offered by computer vision based covid-19 detection are as follows:

- **Improved accuracy:** Computer vision can help to improve the accuracy of diagnosis by identifying patterns in images that are difficult for humans to see.
- **Reduced costs:** Computer vision can help to reduce the costs of diagnosis by automating tasks and freeing up radiologists to focus on more complex tasks.
- **Improved patient care:** Computer vision can help to improve patient care by providing more accurate diagnoses and personalized treatment plans.

All new technologies will also have some challenges associated. Enlisted below are some of these challenges.

- **Data availability:** Computer vision algorithms require large datasets of labelled images to train. However, these datasets can be difficult and expensive to obtain.
- **Accuracy:** Computer vision algorithms are still not as accurate as human radiologists. However, the accuracy of computer vision algorithms is improving all the time.
- **Interpretability:** It can be difficult to interpret the results of computer vision algorithms. This can make it difficult for radiologists to trust the results of computer vision algorithms.

Computer vision is a complex field, but it is also one of the most promising areas of artificial intelligence. As computer vision technology matures, we can expect to see even more amazing and important applications in the future.

2.2.5 Prominent Applications of Computer Vision

Self-driving Cars

Computer vision is used to help self-driving cars navigate the road and avoid obstacles. Many companies are working on self-driving car research. Alphabet which is the parent company of Google has been working with Waymo (Fig. 2.3). Tesla is another important player in this segment. App cab company Uber has also made significant investments in this. Some of the concerns regarding self-driving cars are cost of the cars, safety aspects as well as regulatory aspects of driving these cars.

Figure 2.3 WAYMO self-driving car

Virtual Reality/ Augmented Reality (VR/ AR)

Computer vision is used to create realistic virtual worlds that can be explored by users. Virtual reality creates a completely immersive experience by blocking out the real world and replacing it with a computer-generated environment. Users wear a headset that displays the virtual environment, and they can interact with it using controllers or their own bodies. Augmented reality overlays computer-generated images onto the real world. Users see the real world through a device like a smartphone or headset and they can interact with the computer-generated images by touching them or moving their heads.

An even more exciting concept is **metaverse** which goes beyond the realm of Individual AR/VRs, here multiple people can socialize and participate in the activity in a digital universe. While AR/VR serves as the building blocks, the metaverse needs many more technologies like block chain, cloud computing, etc.

2.3 NATURAL LANGUAGE PROCESSING

Natural language processing (NLP) is a subfield of artificial intelligence that deals with the interaction between computers and human (natural) using languages. More simplistically, this subfield of AI focuses on developing the capability of machines to interpret natural language like a human being. Before we discuss some of the important applications. We will later on see, that the algorithms expect some kind of disciplined data. Data that has a structure, i.e., has a definite set of columns and rows with clear meaning. Text may be very informative, but it lacks structure. Text pre-processing comes to our rescue here.

Text pre-processing is an essential step in natural language processing (NLP) that involves cleaning and preparing text data for analysis and modelling. Some common pre-processing techniques in text include:

- **Lowercasing:** Converting all text to lowercase to ensure consistency and prevent case-related duplication (e.g., 'word' and 'Word').
- **Tokenization:** Splitting text into individual words or tokens, makes it easier to analyse.
- **Stop Word Removal:** Eliminating common, uninformative words (e.g., 'and,' 'the,' 'is') that don't contribute significantly to the analysis.
- **Punctuation Removal:** Stripping text of punctuation marks, which are often irrelevant for many NLP tasks.
- **Numerical and Special Character Removal:** Removing numbers and other non-alphabetic characters, depending on the analysis.
- **Whitespace Trimming:** Removing unnecessary spaces, tabs, or line breaks.
- **Lemmatization and Stemming:** Reducing words to their base or root form to standardize and consolidate related words (e.g., 'running' to 'run').
- **Spell Checking and Correction:** Identifying and rectifying spelling errors in the text.
- **Handling Contractions and Abbreviations:** Expanding contractions (e.g., 'can't' to 'cannot') and standardizing abbreviations.
- **Handling HTML Tags:** Stripping or converting HTML tags in text data.
- **Text Normalization:** Normalizing text data to ensure consistency, for example, converting different date formats to a common standard.

- **Removing or Masking Personal Identifiable Information (PII):** Replacing or removing sensitive information like names, addresses, or social security numbers for privacy and compliance.
- **Removing URLs and Email Addresses:** Eliminating URLs and email addresses that might not be relevant for analysis.
- **Text Segmentation:** Splitting text into segments or paragraphs, depending on the analysis requirements.
- **Sentence and Document Length Normalization:** Ensuring that sentences or documents are of consistent lengths for tasks like text classification.
- **Encoding and Decoding:** Converting text between different character encodings as necessary.

Some of the important tasks in the domain of natural language processing are as follows.

2.3.1 Text Classification

Text classification involves categorizing or assigning predefined labels to a piece of text. For example, sentiment analysis, where text is classified as positive, negative, or neutral based on the expressed sentiment, or topic classification, where text is categorized into specific topics or domains. A typical sentiment analysis AI model will take a piece of text as input and categorize them as positive, negative or neutral. Well, this is for a simplistic understanding, there can be of course many variants and complexities. Following are a couple of examples of sentiment analysis:

- Very good, solid, good balance, comfortable...loved it - > Positive
- Parts are missing and I have given this as a gift and now return window is closed. Very bad experience - > Negative

Topic classification can be used to categorize a news or an article to a given topic, which can help to organize the topics.

2.3.2 Named Entity Recognition (NER)

NER focuses on identifying and extracting specific named entities from text, such as names of persons, organizations, places, dates, or other predefined entities. It helps in information extraction and understanding the context of a text. When a machine is reading a piece of text, NER can go a long way in understanding the text.

Some examples of NER are as follows:

1. 'The President of the United States is Joe Biden.' The NER system would identify 'President' as a title, 'United States' as a location and 'Joe Biden' as a person.
2. 'The company TCS was founded in 1968.' The NER system would identify 'TCS' as an organization and '1968' as a date.
3. **'The meeting is going to be held at 10:**00 AM today.' The NER system would identify '10:00 AM' as a time and 'today' as a date.

NER is a critical component of many NLP systems, such as chatbots, sentiment analysis tools and search engines. It is also used in a variety of other applications, such as:

- **Information extraction:** NER can be used to extract information from text, such as the names of people, organizations, and locations. This information can then be used for a variety of purposes, such as building a knowledge base or generating reports.
- **Machine translation:** It can be used to improve the accuracy of machine translation by identifying named entities in the source text and translating them correctly.
- **Text summarization:** NER can be used to improve the accuracy of text summarization by identifying named entities in the text and summarizing them correctly.

2.3.3 Question Answering

Question Answering (QA) systems aim to automatically answer questions posed in natural language. These systems analyse questions, search for relevant information and generate appropriate answers. QA systems can be based on retrieving answers from a database or using a machine comprehension approach to understand and generate answers.

There are two main types of QA tasks:

1. **Extractive QA:** In extractive QA, the system extracts the answer to the question directly from the text. For example, if the question is 'What is the capital of France?', the system would extract the answer 'Paris' from the text.

2. **Abstractive QA:** In abstractive QA, the system generates an answer to the question that is not directly extracted from the text. For example, if the question is 'What is the meaning of life?', the system might generate an answer that is a philosophical or religious reflection on the meaning of life

Based on the expertise of the QA System, there can be another set of classifications.

- **Closed-domain QA:** These systems focus on answering questions within a specific domain, such as medical QA or legal QA.
- **Open-domain QA:** These systems aim to answer questions on a wide range of topics without a specific domain restriction.
- **Knowledge base QA:** Some QA systems are built on top of structured knowledge bases like DBpedia or Freebase, allowing them to answer fact-based questions more effectively.

QA systems are deployed in a variety of applications, including the following:

- **Search engines:** QA systems can be used to answer questions that users ask in search queries.
- **Virtual assistants:** QA systems can be used to answer questions that users ask virtual assistants, such as Amazon Alexa or Google Assistant.
- **Knowledge bases:** QA systems can be used to answer questions that users ask about a knowledge base, such as a database of product information or a database of medical information.

QA systems can very naturally be used in Quora. It might be of interest to the readers that a Kaggle Competition (https://www.kaggle.com/c/quora-question-pairs) was launched around 2017. Here, the task that was given to solve was identifying question pairs which are similar. In fact, the latest favourite of everyone, ChatGPT is also an example of a QA System.

2.3.4 Machine Translation

Machine translation focuses on automatically translating text or speech from one language to another. It involves understanding the source language, generating an intermediate representation and producing the translated output in the target language. The machine translation is done mainly following three approaches:

1. **Rule-based MT:** Rule-based MT systems use a set of rules to translate text. The rules are typically based on the grammar of the two languages and the context of the words in the text.
2. **Statistical MT:** Statistical MT systems use statistical methods to translate text. The systems learn to translate text by analysing a large dataset of parallel text.
3. **Neural MT/AI-based MT:** Neural MT/AI Based MT systems use artificial neural networks to translate text. Artificial neural networks are a type of machine learning model that has a special ability to learn complex patterns in data.

There is a lot of focus from the Central Government on machine translation from one Indian language to another. Interested readers may visit www.bhasini.gov.in for more details. The objective is digital empowerment and digital inclusion by making the ecosystem available in most of the Indian languages. One notable initiative in this regard is named Bhasha Daan. Here, all citizens can contribute to building an open-source dataset.

Google Translate has been there for long time now and is supporting more than 120+ languages worldwide.

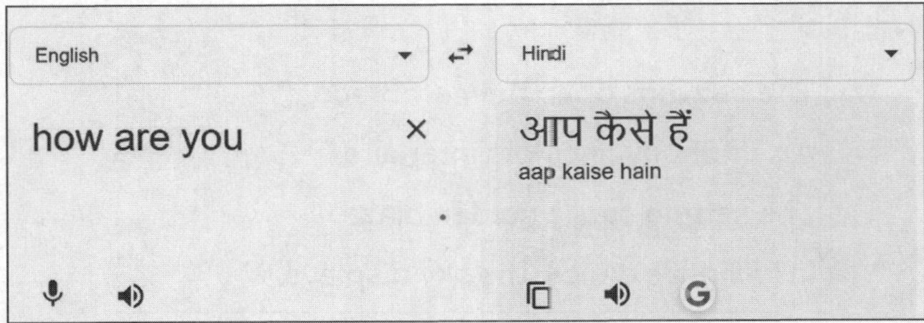

Figure 2.4 Machine Translation using Google Translator

Well, the above example (Fig. 2.4) is at a toy scale. Google Translator is doing that translation for 24 × 7 over numerous examples and are collecting active feedback and improving itself further. But genuinely language is a very complex thing and the current state of the art is indeed good but far from perfect.

2.3.5 Text Generation

Text generation involves creating coherent and meaningful text in natural language. This can range from simple tasks like generating product descriptions to more complex applications like story generation or language modelling. Similar, to other NLP tasks, the primary three approaches of text generation are (i) rule based methods (ii) statistical methods and (iii) neural or AI-based methods.

Text generation has a variety of use cases. Couple of crucial ones are mentioned below:

- **Chatbots:** Chatbots are AI-based programs that can simulate conversations with humans. Chatbots are often employed in customer service applications, where they can answer questions and provide support to customers.
- **Content creation:** Text generation capability can be leveraged to create content, such as articles, blog posts and social media posts. This can be useful for businesses that want to create content that is engaging and informative. Shown in Fig. 2.5, is a conversation with ChatGPT, where the transformer-based language engine generates poetry like a professional.

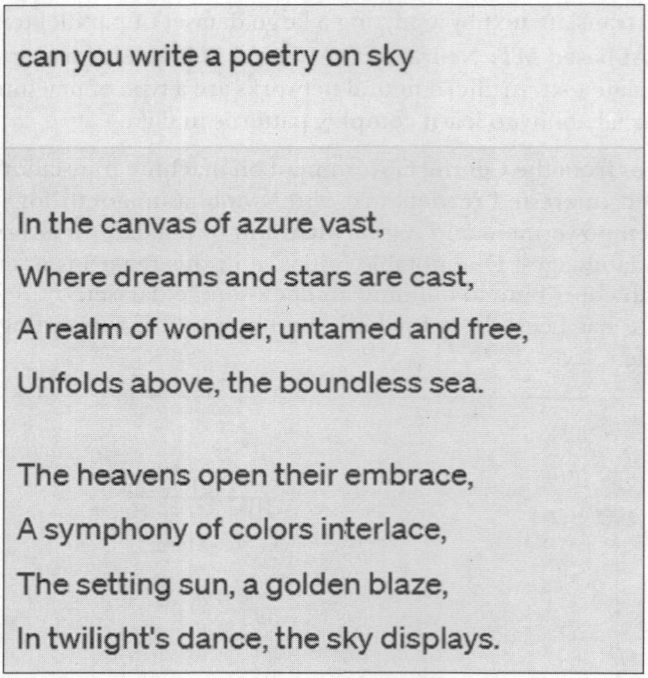

Figure 2.5 Content Generation Example using a Conversational Agent

2.3.6 Text Summarization

Text summarization aims to generate concise summaries of longer texts, condensing the main points or key information. Summarization techniques can be extractive (selecting and rearranging important sentences) or abstractive (generating new sentences to capture the essence of the text).

There are two main approaches to text summarization which is discussed hereunder.

Extractive Summarization

Extractive summarization involves selecting and extracting sentences or phrases directly from the original text to form the summary. The extraction is typically based on various criteria, such as sentence importance, relevance to the main topic and coherence with other selected sentences. The advantage of extractive summarization is that it maintains the exact wording and context from the original text, but it can sometimes result in less fluency and coherence in the summary.

Abstractive Summarization

Abstractive summarization, on the other hand, involves generating new sentences that may not appear in the original text to form the summary. In this approach, the model understands the content of the input text and rephrases or rewrites the information in a concise and coherent manner to create a summary. Abstractive summarization can produce more fluent and human-like summaries, but it is generally more challenging due to the need for language generation and maintaining the intended meaning.

Text summarization methods can use various techniques, including statistical approaches, rule-based methods, and machine learning, similar to the case of machine translation.

Applications of text summarization include:

- **News summarization:** Automatically generating concise summaries of news articles to provide users with quick and informative overviews.
- **Document summarization:** Condensing lengthy documents, research papers, or reports into shorter summaries for easy understanding and reference.
- **Social media summarization:** Summarizing long social media posts or discussions to highlight the main points or sentiments.
- **Content curation:** Summarizing multiple pieces of content to provide users with curated and relevant information.

Text summarization plays a crucial role in helping users navigate through vast amounts of textual data efficiently and can significantly save time and effort in information retrieval and understanding.

2.4 MACHINE LEARNING

Machine learning is a sub-discipline of artificial intelligence, which depends heavily on data. In a typical scenario, when we write a computer program for let's say finding whether a number is prime or not, we give an input, an integer in this case to the computer program and the program gives an output either true or false. So, if an input 7 is given to the program, the output will be true and for any composite number let's say 20, the output will be false. The logic of checking primality needs to be coded in some programming language. The basic logic may be something like this, for the number we are testing, we will try to divide it by all numbers between 1 and the number itself. If it is divisible by anyone of them, then it is not a prime number. The logic we wrote in natural language just now, can be written more concisely. Such a representation of the logic, which is human readable is called an algorithm, which is often the first step of writing a program.

One other thing, that needs to be discussed in this context is that machines do not understand human languages, they can only understand strings of 0s and 1s. These are called binary strings. However, humans cannot express logic in terms of 0s and 1s. There are several programming languages which has their own grammar. If it is close to machine language then it is a low-level language and if it is close to natural language then it is a high-level language.

A program like the above-mentioned can easily be written in a high-level language. Let's look at the problem statement a little more analytically. Here the objective of the program is clear and the steps to achieve the same are also clear. There are several toy problems like this:

1. Swapping the value of two variables
2. Find maximum of three numbers
3. Finding the factorial of a given integer
4. Find the greatest common divisor of two numbers
5. Find whether a particular year is a leap year or not

The programming paradigm that we discussed works fine for all five problems. However, let's look at some other types of problems.

1. A customer is approaching a bank for a loan, looking at his demographic information, and past financial record, can we say whether he is going to default or not.
2. Classifying an email as a normal email or spam email.
3. Predicting whether tomorrow there will be rainfall or not

There is a clear difference here. The problem definition is clear, but the exact steps to find the same is not clear. These are the problems that are attempted to be solved using **machine learning.** Typically, this is learned from past examples. For the first problem, past customers data will be looked at. Table 2.1 enlists some sample customers.

Table 2.1 Loan Default Sample Data

Cid	Gender	Age	Marital Status	Monthly Salary	Loan Default
1	M	30	Single	30000	Y
2	F	41	Married	40000	N
3	F	31	Married	25000	N
4	M	27	Single	60000	Y
5	M	52	Married	70000	N

Gender, age, marital status and monthly salary are called input features or attributes. Loan default is called the output attribute or target attribute, also called the label. A machine learning program looks at the data and learns patterns on its own to associate the input with the output. These past examples serve as the training set for the machine learning program, which now develops a capability, given the attributes of a customer it can predict whether the customer will default or not. Such type of learning through examples is called inductive learning.

If we attempt to categorize, the machine learning tasks, it will look like Fig. 2.6.

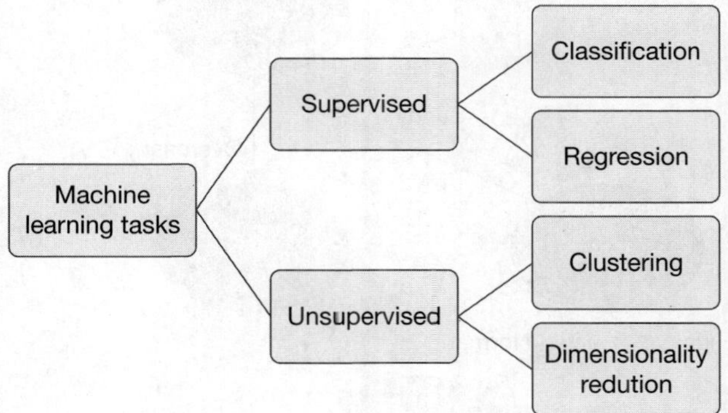

Figure 2.6 Categories of Machine Learning Tasks

This is a very simplistic representation. Many more categories can be added as a refinement to the above classification. However, it serves as a baseline.

2.4.1 Classification

Classification is the task of categorizing input data into predefined classes or categories. The model is trained on labelled data, where each data point is associated with a specific class label.

Examples of classification tasks include spam detection, sentiment analysis, image classification and text categorization.

Figure 2.7 shows a sample dataset of playing golf.

1	Outlook	Temp	Humidity	Windy	Play Golf
2	Rainy	Hot	High	FALSE	No
3	Rainy	Hot	High	TRUE	No
4	Overcast	Hot	High	FALSE	Yes
5	Sunny	Mild	High	FALSE	Yes
6	Sunny	Cool	Normal	FALSE	Yes
7	Sunny	Cool	Normal	TRUE	No
8	Overcast	Cool	Normal	TRUE	Yes

Figure 2.7 Golf Dataset

This is a popular playing golf dataset. Outlook, temperature, humidity and windiness are the input attributes and playing golf is the target attribute. So, the task to be done is to develop a model, based on the given data about a particular day, whether it will be good for playing golf. There are several classification algorithms. One such algorithm is also called decision tree. This is shown in Figure 2.8.

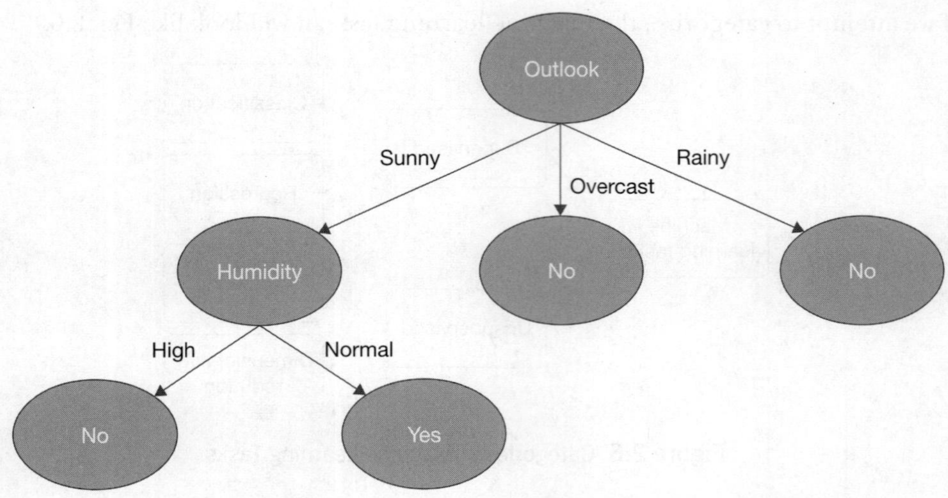

Figure 2.8 Partially Constructed Decision Tree on Golf Dataset

So, if on a particular day, the outlook is sunny and the humidity is normal, irrespective of other attributes, the decision is to play (YES). You might observe that the above representation does not look like a tree, well it's an inverted tree. The node on the top is called the root node and the decision nodes i.e., the last level of nodes are called leaf nodes.

These decision trees are used in two phases. In phase 1, they learn from previous data and in phase 2, they can predict. Let's take some other examples. Let's say, we are working with a problem of predicting whether a particular employee will stay in the organization or leave the organization. So, in phase 1, a decision tree will be constructed from past data (previous employee records their demographic and behavioural data and a target field, whether they stayed or left), and then in phase 2, when a new employee's record is given, using the decision tree, it can be predicted whether the employee will be staying or leaving.

Hence, the decision tree can work as a classification model. This model is called a white box model as the reason for reaching the final decision is amply clear. As we move to a more complex model, this becomes opaque and such models are called black box models.

There are several other models for classification namely k nearest neighbour (kNN), Naïve Bayes (based on conditional probability), logistic regression, support vector machine, etc. There is also a neural network-based model named as multi-layer perceptron, which we will discuss later.

2.4.2 Regression

Regression is the task of predicting continuous numerical values based on input data. The model learns the relationship between the input features and the target numerical value during training. Regression is commonly used for tasks such as predicting house prices, sales forecasting and time series analysis. The easiest variation of regression is when we have one input variable and one output variable. The input variable is often called the regressor or predictor. The simplest of the models is called a linear regression. Let's say, we are trying to find out that for a science graduate, based on their CGPA, what should be their in-hand salary. Like, the classification problem, there will be past data about the students.

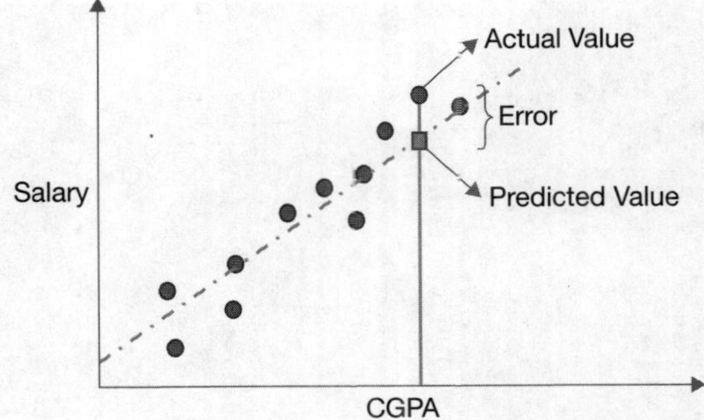

Figure 2.9 Scatter Plot and Regression line Salary vs CGPA

In Fig. 2.9, CPGA, the input variable is often called the independent variable is drawn along the X-axis and Salary the dependent variable is drawn along the Y-axis. Each circle above represents data about a past student. Based on these circles, a line has been drawn such that **collectively the circles are as close as possible to the line.** The previous sentence is worth a lot of attention. We can draw several alternative lines, which are close to the circles, however, we are interested about the line, which is closest to all. Some curious readers may want to quantify closeness.

Well, if you look thoroughly for one of the circles at the extreme right, a straight-line projection is shown from the circle to the regression line. The value on the line is the predicted value and the circle is the actual value. Error is given as Actual Value – Predicted Value. When, there is only one input variable it is called simple linear regression and if there are more than one input variable, we call this approach multiple linear regression. These are linear models because the relationship between the dependent and independent is assumed to be linear (the model is a straight line). Instead of a line, it can be a more complex one like a polynomial.

It might be noted that both classification and regression are examples of supervised learning. Who gives the supervision? The past experience embodied in terms of labelled data gives supervision.

2.4.3 Clustering

Clustering is an unsupervised learning task where the goal is to group similar data points together based on their similarity. Data points within the same cluster are more similar to each other than to those data points in other clusters. Clustering is a more difficult task than supervised learning as there is no past data with labels. Hence, the algorithms have to learn on their own.

Figure 2.10 illustrates the dilemma. If someone asks us to group the building block pieces, they can be grouped either by shape or by size or even by colour. Clustering algorithms can be classified further. One way of classification is exclusive and nonexclusive clustering methods. In an exclusive algorithm, a single point can belong to only one cluster and in case of non-exclusive algorithm, a particular point or observation can go to multiple clusters. For example, if one cluster stands for aquatic animals and another cluster stands for land-living animals. A turtle or a frog will belong to both clusters.

Figure 2.10 Kid's building Blocks

One important clustering algorithm is called *k*-means. When applied to a dataset this algorithm can find *k*-clusters. The detailed working is beyond the scope of the current book.

2.4.4 Dimensionality Reduction

Dimensionality reduction is the process of reducing the number of features or variables in the data while preserving its important structure. This helps in visualizing and understanding high-dimensional data and can improve the performance of machine learning models.

Imagine you have a bunch of data, like a list of people's heights, weights, and ages. Now, think of this data as points in a three-dimensional space where each axis represents one of these measurements. It's like a 3D graph where each point represents a person. Dimensionality reduction techniques are a way to look at this data and figure out what's most important. Imagine you want to simplify this 3D graph into a 2D graph but keep as much information as possible. The motivation for dimensionality reduction techniques is that.

Principal Component Analysis (PCA) and t-distributed Stochastic Neighbour Embedding (t-SNE) are popular techniques used for dimensionality reduction. There are many benefits of dimensionality reduction. Firstly, with reduced dimensions, there is less requirement for storage. Then, the machine learning models that are being built also require less time to be constructed and finally, with a smaller number of features, it becomes much more explainable.

2.4.5 Recommendation Systems

As we live our lives, we consume, use or do a variety of things. That may be home appliances like TV, Fridge, Washing Machine, etc., that can be buying groceries, maybe planning a trip, going for a family dinner, reading a book, watching a movie or applying for a job. You just name it. For all these actions, there are a variety of choices available and that's where the recommender system comes in. It keeps track of your usage, your choices and gives you a recommendation. Maybe you order Chinese food on weekends, so based on that a recommendation comes from the food delivery app, when a new Chinese eatery has opened up and gives a lot of discounts.

DO YOU KNOW

There is an interesting story of how this all started. Netflix (one of the largest subscriptions based streaming service provider), in the year 2006, announced a one Mn $ prize money to any ML researcher, group of researchers who can better their recommendation. This kind of money generated a lot of interest and actually many algorithms got invented for the same. The competition lasted for three long years.

More formally, recommendation systems aim to predict and suggest items or content that a user might be interested in. Two principal categories of recommendation are (a) Content based recommendation (b) Collaborative Recommendation or Filtering.

Content-based recommendation is based on the static characteristics of the item. Taking the inspiration from Netflix, continuing with the selection of movie examples. Content-based recommendation will compare two movies based on their short description, genre, director and actor information etc.

Collaborative filtering is more on the basis of the dynamic characteristics of an entity. Let's say, you like action movies and love thrillers and do not enjoy the romcoms. This technique will quickly identify users who are similar to you and if this group of users thinks highly of a movie (given a high rating) and this is not yet watched by you, it will be recommended to you. This type of filtering which is based on similar types of users is called user-based collaborative filtering.

There is another type. Let's say, you liked movie 1, so similar movies will be found out. How this similarity will be determined? If the same set of users give the same kind of rating to two movies, then they are similar, let's say movie 1 and movie 2 are two such movies. Now, if movie 2 is not watched by you, then it will be recommended to you. Such a type of collaborative filtering is called item-based filtering.

There are some interesting problems as far as recommender systems are concerned. Let's discuss a couple of them here.

Long Tail Problem

In any digital marketplace, there are a lot of merchandise. The count of such products may easily run into thousands. While some of the items are popular and rated by a lot of people, some items are consumed and rated less.

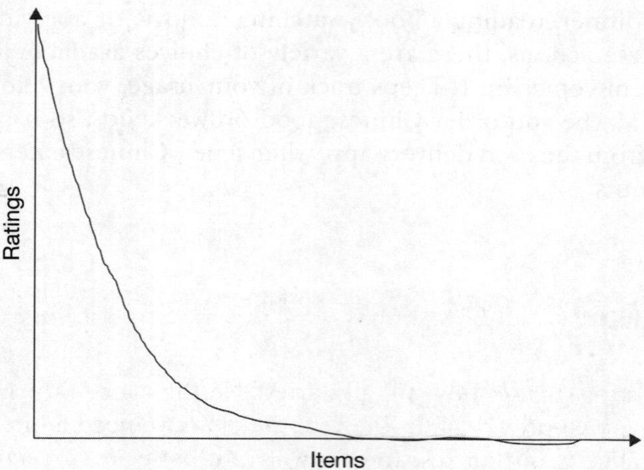

Figure 2.11 Long Tail Distribution of Ratings

The items which are less rated appear at the tail of the distribution and this tail is quite long. The Long tail distribution of product ratings is shown in Figure 2.11. Recommending these items becomes quite difficult.

Cold Start

This is a problem when a new item is introduced. There are no ratings available for this item. Hence, recommending such an item also becomes difficult.

2.4.6 Anomaly Detection

Anomaly detection involves identifying unusual or abnormal data points that deviate significantly from the normal pattern. It is used in fraud detection, fault detection and cybersecurity. This is also referred to as outlier detection, novelty detection etc. Some important use cases of anomaly detection are fraud detection, intrusion detection, fake news detection, etc.

Let's further elaborate on this with an example. Generally, every person has a regular way of doing his or her spending. For example for Arnab, it may be ATM withdrawals on Monday, Grocery shopping on weekends, Movies or fine dining maybe once a month, some medical expenses, fees for school, etc. Any transaction which is high value and does not fit in this regular pattern (anomaly) may be thought of as suspicious or fraud.

2.4.7 Deep Learning as a Special Case of Machine Learning

As mentioned earlier, the grand aim of AI is to create an artificial human brain. The basic unit of the human brain is a biological. Hence, the starting step of AI will be to create artificial neurons. Late, 1950's the perceptron model was popularized, which had the capability of simple decision-making. Figure 2.12 shows a biological neuron with its artificial counterpart.

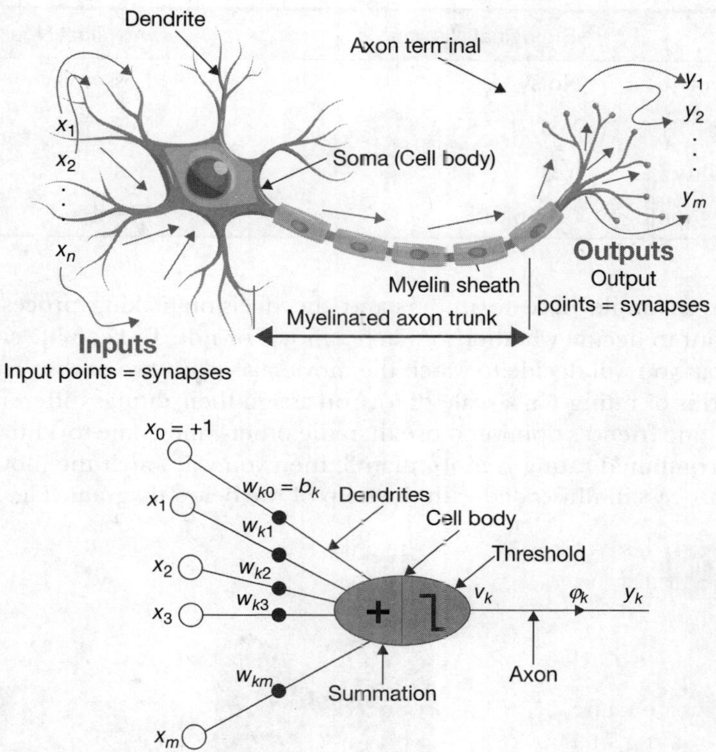

Figure 2.12 Biological neuron with its artificial counterpart

A biological neuron takes inputs through synapses from other neurons. They are aggregated and the summed input may trigger the neuron, in which case the neuron fires and a signal flows through the axon to the dendrites. This simple artificial neural network was also called perceptron. Apparently, these could be used, to simulate a simple decision-making process (Binary Choices).

The below table furnishes the differences between artificial neuron and biological neurons.

Property	Biological Neuron	Artificial Neuron
Composition	Living cells	Computer code
Inputs	Many (dendrites)	Few
Outputs	One (axon)	Many
Signals	Electrical spikes (action potentials)	Continuous values
Speed	Slow (up to 200 Hz)	Fast (millions of Hz)
Energy efficiency	Efficient	Inefficient

(Continued)

Property	Biological Neuron	Artificial Neuron
Noise level	Noisy	Less noisy
Learning	Yes	Yes
Adaptability	Yes	Yes
Network topology	Complex	Flexible

Let's look at an example here to understand the decision-making process of a neuron. Let's say, you want to decide whether to watch a movie or not. To keep life simple, here our assumption is that you will decide to watch the movie, based on two of your friend's opinions expressed in terms of rating (In a scale of 5). You assign their ratings different weights (well you may rely on one friend's opinion more than the other) and come to a final rating (aggregation). If this combined rating is more than 3, then you will watch the movie (the neuron will fire). This process in illustrated with the help of the below diagram (Fig. 2.13).

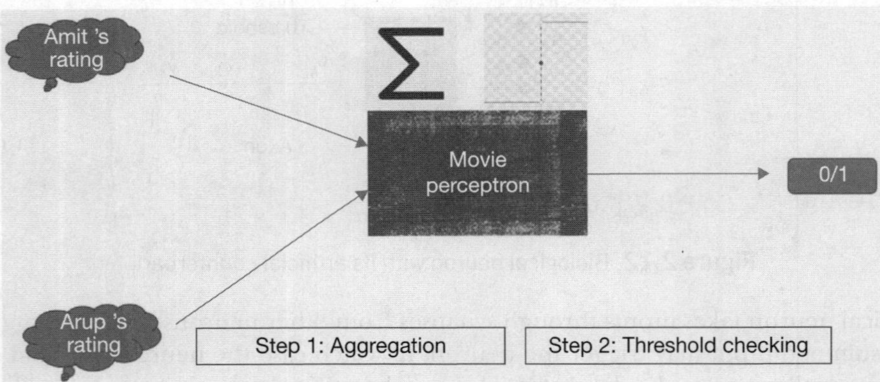

Figure 2.13 Step by step decision making by ANN

Let's say Amit rates the movie 3.5 and Arup rates it 2.5. For some reason, we give a weightage of .6 to Amit and then .4 to Arup. Then the combined rating is 3.5 * .6 + 2.5 * 4 = 2.1 + 1 = 3.1. The threshold as mentioned is 3, hence you decide to watch the movie. Well, without making things two complicated, let's try to put this mathematically. Let, the rating of Amit is represented by R_1 and the weightage to his rating is represented by W_1.

On a similar note, R_2 and W_2 represent Arup's rating and weightage of rating respectively. The value or threshold which is 3 in this case is presented by W_3. Hence, if the following equation gives a positive value, then, you watch the movie, else you do not.

$$w_1 * x_1 + w_2 * x_2 - w_3 \geq 0$$

Well, if you look at the equality part carefully and concentrate on $w_1 * x_1 + w_2 * x_2 - w_3 = 0$. This actually represents, the equation of a line. The inequality also serves as an activation function.

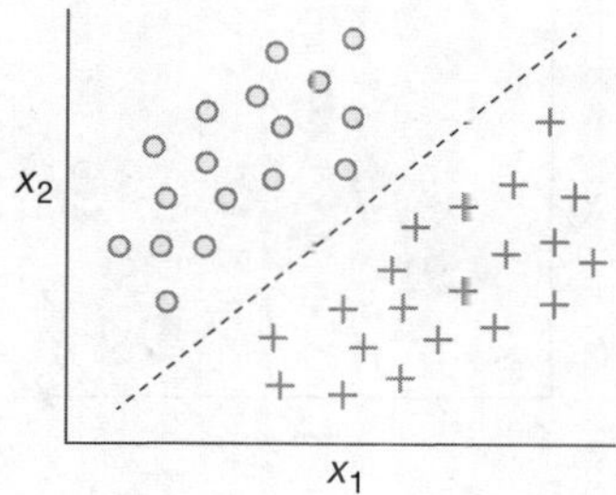

Figure 2.14 Linearly separable classes

If we look at Fig. 2.14, the circles and plus symbols represent individual movies. If the ratings of the movies are on the right side of the dotted line, then you will watch the movie, else you won't watch it. The dotted line in this case is called the decision boundary.

The Perceptron algorithm boosted the capability of perceptron further. Given a problem and a few past movies (rating of Amit and Arup for those movies and your choice of watching the movie or otherwise after knowing your friends' rating is available) data is available, then from the data itself, the algorithm can learn the values of w_1, w_2 and w_3. This makes the perceptron a general decision-making tool.

After the initial euphoria was over, one scepticism, that hit the perceptron community was that it will not be suitable for complex problems. They observed, that perceptron can only work when these two classes (Movies I am going to watch and movies I am not going to watch) are separable by a straight line.

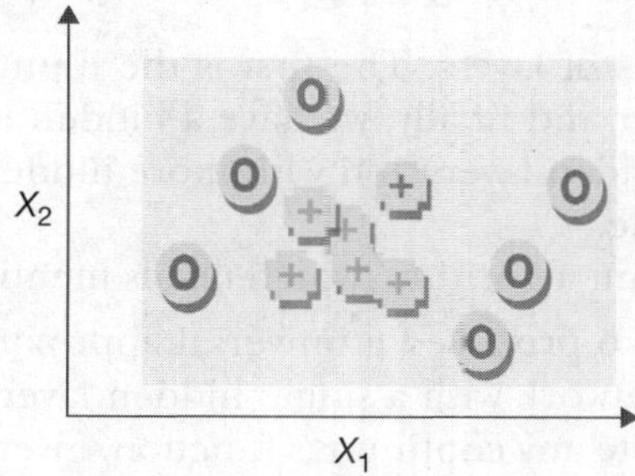

Figure 2.15 Classes not non-linearly separable

For example, in Fig. 2.15, the circled movies and the plus sign movies cannot be separated by a line. The problem described may look really trivial, however, the answer remained elusive for quite some time, ushering one of the AI Winters (a period, when there was a lack of optimism, research and funding in anything directly or even remotely linked with AI).

As more research continued, scientists realised while this may not be achievable by a single line, if multiple lines can be employed, a complex decision boundary can be achieved. Such a probable solution is depicted below in the diagram (Fig 2.16).

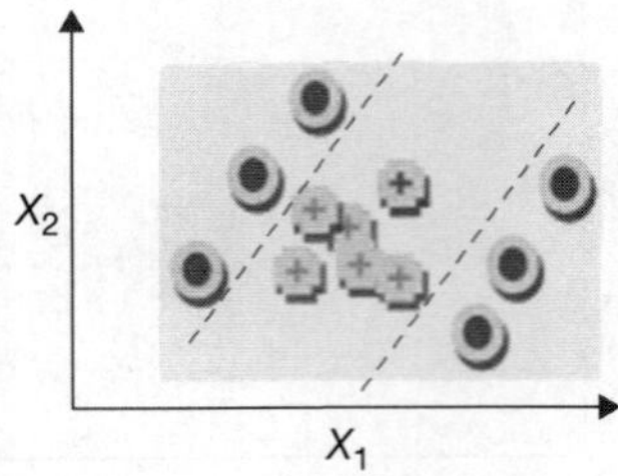

Figure 2.16 Classes not non-linearly separable, separated using two lines

Hence, two perceptrons can be used to draw two lines and then another unit can check whether the movie or observations are falling within the two lines or outside of any of the lines. This led to the theory of the Multi-Layer Perceptron. Figure 2.17 represents such a setup.

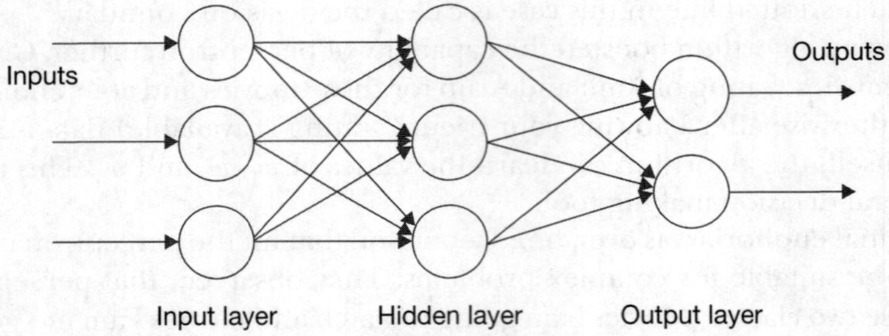

Figure 2.17 Building blocks of a multi-layer perceptorn

Typically, there are three types of layers. The first is the input layer, where it takes inputs. Then, there is the output layer and finally, we have a hidden layer in between. It turns out that, there can be multiple hidden layers and with more hidden layers, the resultant neural network becomes more capable.

There is one more important algorithm, which needs mention.

- In 1989, George Cybenko proposed a universal approximation theorem that showed a feedforward neural network with a single hidden layer and a continuous activation function can approximate any continuous function, given enough hidden units.
- Shortly after Cybenko's work, Kurt Hornik, Maxwell Stinchcombe and Halbert White showed that the same result holds for feedforward neural networks with multiple hidden layers. They also showed that the specific choice of activation function is not important, as long as it is continuous.

The universal approximation theorem is one of the most important results in the field of artificial intelligence. It shows that neural networks are capable of learning any complex pattern in data, as long as the pattern is continuous. Networks, with 2 or more hidden layers are called deep neural networks.

While the concept of multi-layer perceptron came around the 1980s, it took rigorous research for around 25–30 years, to start reaching the potential of these networks. There are three gentlemen namely Geoffrey Hinton, Yoshua Bengio and Yann Lecun, who kept up their research. They are rightly called the fathers of deep learning. In 2018, they were awarded with the Turing Award (Highest Honour in Computer Science).

One question, that will surely come to your mind is what changed in the course of time, to make deep learning models successful. Below is a summary of the same.

GPU Acceleration

The availability of powerful Graphics Processing Units (GPUs) significantly accelerated the training of deep neural networks, making it feasible to work with large-scale models and datasets. Without too much technical nitty-gritty, GPU may be assumed to be like a supercharged, specialized computer chip inside your computer that's really good at handling tasks related to graphics and calculations. It's like having a separate, powerful brain just for handling images, videos, and complex math problems.

Access to Big Data

The proliferation of digital data and the internet have provided an abundance of labelled and unlabelled datasets, which is essential for training deep learning models effectively. Compared to machine learning models, the deep learning models have more parameters. More the parameters, more is the requirement of data.

Advancements in Algorithms

Breakthroughs in neural network architectures (e.g., convolutional neural networks and transformers) and optimization techniques (e.g., stochastic gradient descent) have greatly improved the performance of deep learning models, enabling them to tackle complex tasks.

- Deep learning has achieved remarkable success in various domains, including computer vision, natural language processing, and speech recognition.
- Convolutional Neural Networks (CNNs) excel in image-related tasks, such as object detection and image classification. CNN is specialized for images. It turns out that an image is formed in our visual cortex in quite a layered manner. (In 1981, Hubel and Wiesel, got the Nobel prize for explaining the visual processing of the brain).
- Recurrent Neural Networks (RNNs) are well-suited for sequential data and time series analysis. Some examples of sequence data are language, weather data, stock market data, data from different sensors, etc.
- Long Short-Term Memory (LSTM) networks are a specialized form of RNNs designed to capture long-range dependencies in sequences.
- Transformers have revolutionized natural language processing and are the foundation for many state-of-the-art language models. Current sensations like ChatGPT, Bard etc., are based on transformer architecture.
- Transfer learning and pre-trained models have become common techniques for leveraging existing deep learning knowledge. This needs further explanation. Knowledge acquired from one task is very conveniently used by humans in another task. Knowledge of walking can be used in running or climbing stairs. Knowledge of vocabulary can be used for playing scrabble or writing an essay. Knowledge of playing badminton can be used in playing table tennis. Often, it also happens because of your already acquired knowledge of badminton (pre-training), it can be transferred to playing table tennis (With some adjustments, known as fine-tuning). This is possible for deep learning as well. A model trained for one image classification task can be used for another image classification task with minor adjustments. It may be noted that transfer learning was not feasible in machine learning models.

Let's look at some of the important deep learning architectures.

Model	Description	Use cases
Convolutional Neural Networks (CNNs)	CNNs are a type of deep learning model that is particularly well-suited for image recognition and classification tasks. They work by extracting features from images using a series of convolutional and pooling layers.	CNNs are used in a wide variety of applications, including image classification, object detection, and facial recognition. For example, CNNs are used to power the image search and photo tagging features in Google Photos and Facebook.
Recurrent Neural Networks (RNNs)	RNNs are a type of deep learning model that is well-suited for sequential data, such as text and audio. They work by processing the data one sequence element at a time and maintaining a hidden state that captures the information from the previous elements in the sequence.	RNNs are used in a variety of applications, including natural language processing, machine translation, and speech recognition. For example, RNNs are used to power the Google Translate and Siri features.
Long Short-Term Memory (LSTM)	LSTMs are a type of RNN that is particularly well-suited for tasks that require long-range dependencies, such as machine translation and speech recognition. LSTMs work by using a special type of memory cell to store information from the previous elements in the sequence.	LSTMs are used in a variety of applications, including natural language processing, machine translation, and speech recognition. For example, LSTMs are used to power the Google Translate and Siri features.
Generative Adversarial Networks (GANs)	GANs are a type of deep learning model that consists of two competing networks: a generator network and a discriminator network. The generator network tries to generate realistic data, while the discriminator network tries to distinguish between real and generated data.	GANs are used in a variety of applications, including image generation, text generation, and music generation. For example, GANs are used to power the DeepDream feature in Google Photos, which can generate psychedelic images from ordinary photos.
Transformers	Transformers are a type of deep learning model that is particularly well-suited for natural language processing tasks. Transformers work by using a self-attention mechanism to learn long-range dependencies in the data.	Transformers are used in a variety of natural language processing tasks, including machine translation, text summarization, and question-answering. For example, Transformers are used to power the Google Translate and Bard features.
Autoencoders	Autoencoders are a type of deep learning model that are trained to reconstruct their input data. They consist of two parts: an encoder and a decoder. The encoder learns to compress the input data into a latent representation, and the decoder learns to reconstruct the input data from the latent representation.	Autoencoders are used in a variety of applications, including anomaly detection, dimensionality reduction, and data denoising. For example, autoencoders are used to detect fraudulent credit card transactions and to reduce the dimensionality of images before classification.

2.4.8 Reinforcement Learning

Reinforcement learning is a type of machine learning where an agent learns to behave in an environment by trial and error. The agent receives rewards for taking actions that lead to desired outcomes, and punishments for taking actions that lead to undesired outcomes. Over time, the agent learns to take actions that maximize the expected reward. Reinforcement learning is a powerful technique that can be used to solve a wide variety of problems, such as playing games, controlling robots and making financial decisions.

Some key concepts of reinforcement learning are as follows:

- **Agent:** The agent/a special software program is the entity that learns to behave in the environment.
- **Environment:** The environment is the world in which the agent acts.
- **State:** The state is the current condition of the environment.
- **Action:** An action is something that the agent can do.
- **Reward:** A reward is a signal that the agent receives for taking an action that leads to a desired outcome.
- **Punishment:** A punishment is a signal that the agent receives for taking an action that leads to an undesired outcome.
- **Policy:** A policy is a rule that the agent uses to select actions.
- **Value function:** A value function is a function that estimates the expected reward for taking an action in a given state.

Below is a simple example of how reinforcement learning can be used to train a robot to walk:

1. The robot starts in a random position.
2. The robot takes an action, such as moving its left leg forward.
3. The environment provides the robot with a reward or penalty, depending on the outcome of the action. If the robot takes a step forward without falling down, it receives a reward. If the robot falls down, it receives a penalty.

The robot repeats steps 2 and 3 until it reaches its goal, or until it runs out of time.

Multi-Armed Bandit

Multi-Armed Bandit is a very fundamental problem as far as reinforcement learning is concerned. A multi-arm slot machine is shown in Fig. 2.18.

Figure 2.18 Multi-Armed Bandit

Let's quickly understand what is a single armed bandit. It is a slot machine with a single lever deployed at casinos or often at fairs. The setup is simple, there is a lever, which needs to be pulled after inserting coin or money. When the lever is pulled, some numbers or symbols are displayed. Some of the combinations of symbols are winning combinations, when the player gets some reward in the form of money or tokens. However, in most cases the player loses and the machine ends up robbing money from the players, hence the name bandit. When there are multiple slot machines, this is called as multi-armed bandit problem. Different slot machines will have different probabilities and certainly, the player would want to play in the machine which has the highest winning chances or probability.

Here's a short step-by-step description of how a multi-armed bandit problem works.

Setup

Imagine you are in a casino standing in front of several slot machines (the 'one-armed bandits'). Each slot machine has its own unknown probability of giving you a reward when you pull its lever. Your goal is to maximize your total reward by choosing which slot machines to play and how often to play them.

Exploration vs. Exploitation

In the beginning, you don't know which slot machine is the most rewarding. So, you face a trade-off between exploring (trying out different machines to learn their winning chances) and exploiting (choosing the machine that seems to be the best based on your current knowledge).

Strategy

You start with a strategy or policy. This strategy dictates how you decide which machine to play at each step.

There are different strategies you can use, such as:

- **Greedy strategy:** Always play the machine that has the highest estimated reward so far.
- **Epsilon-greedy strategy:** Most of the time, play the machine with the highest estimated reward, but occasionally explore other machines randomly (with probability ε). ε is a value between 0 to 1. Value of .1 means 10% exploration, .2 means 20% exploration. These are typical values of ε. However, the value of ε may be changed based on the type of the problem.
- **UCB (Upper Confidence Bound):** Choose machines based on the uncertainty in their estimated rewards. As we pull different levers, we win at different rates. The average reward received for each slot machine can be calculated very easily. UCB also considers the number of trials that were considered, while calculating the average return. If the average return is calculated on the basis of lees no. of trials, then uncertainty is greater. Machines with high uncertainty are explored more.
- **Thompson sampling:** Sample from a probability distribution for each machine and play the machine with the highest sampled value. Here, the experiment starts with different prior assumptions about the reward/ winning chances associated with the machines.

Repeated Actions

We start pulling the levers of the chosen machines according to our strategy. After each pull, we observe whether we win a reward or not.

Learning

As we play more, we update our estimates of each machine's reward probabilities based on the outcomes we have observed. The goal is to improve the accuracy of your estimates to make better decisions.

Iteration

We repeat steps 3–5 for a certain number of rounds or until we achieve a desired level of reward.

Adaptation

Over time, our strategy adapts as we learn more about the machines' reward probabilities, shifting the balance between exploration and exploitation.

The multi-armed bandit problem is a fundamental concept in reinforcement learning, decision-making under uncertainty, and optimization. It has applications in various fields, including online advertising, clinical trials, and recommendation systems, where you need to make choices in an uncertain environment.

Q-Learning

Q-Learning is another important paradigm in reinforcement learning. Multiple arm bandit deals with uncertain environments, but there are some tasks which are continuous in nature. In Q-learning, an agent learns a policy to maximize cumulative rewards by interacting with an environment with multiple states, actions, and rewards. The agent interacts with an environment characterized by states and actions. It learns a Q-value for each state-action pair, allowing it to make sequential decisions based on its current state and knowledge of the Q-values for available actions. The learning process is typically more involved, as the agent must explore, learn, and exploit to optimize its decisions over time. Q-learning is well-suited for problems that involve sequential decision-making, such as game playing, robotics, and control systems. It can handle more complex environments with state transitions. Markov Decision Process (MDP) is linked with Q Learning.

A Markov Decision Process (MDP) is a mathematical framework used to model decision-making problems in a stochastic environment. Here's a brief explanation of an MDP:

- **States (S):** An MDP represents a system or environment that can be in various states. These states could be discrete (e.g., grid cells in a game) or continuous (e.g., real numbers in a control system).
- **Actions (A):** At each state, an agent can take a set of possible actions. These actions are the choices available to the agent to influence the environment.
- **Transitions (T):** When the agent takes an action in a particular state, the system transitions to a new state. The transition function, T, describes the probabilities of moving from one state to another based on the chosen action.

- **Rewards (R):** At each state, the agent receives a numerical reward, which represents the immediate benefit or cost of being in that state. The reward function, R, defines these rewards.
- **Policy (π):** A policy is a strategy or plan that guides the agent's decision-making. It specifies which action to take in each state.
- **Goal:** The objective in an MDP is typically to find an optimal policy that maximizes the cumulative reward over time.

2.5 ROBOTICS

Robotics is the interdisciplinary field of engineering, science, and technology that deals with the design, construction, operation, and application of robots. Robots are machines that can sense their environment and take actions to achieve specific goals. They can be used for a variety of tasks, such as manufacturing, healthcare, and customer service. The field of robotics is constantly evolving, as new technologies are developed to make robots more intelligent, autonomous, and adaptable.

2.5.1 Component of a Robot

Let us quickly look at the basic components of a robot.

- **Frame:** The frame provides the robot with its structure and shape. It is usually made of metal or plastic.
- **Actuators:** Actuators are the components that move the robot. They can be motors, pneumatic cylinders, or hydraulic cylinders.
- **Sensors:** Sensors allow the robot to interact with its environment. They can detect things like light, heat, touch, and sound. Common types of sensors include:
  - **Proximity Sensors:** Detect the presence of nearby objects or obstacles.
  - **Light Sensors:** Measure ambient light levels.
  - **Temperature Sensors:** Measure the temperature of the environment.
  - **IR (Infrared) Sensors:** Detect infrared radiation and can be used for distance measurement.
- **Controller:** The controller is the brain of the robot. The control system of a robot includes the hardware and software that process sensor data and make decisions on how the robot should respond. It can range from simple microcontrollers to complex computing systems. The control system determines how the actuators respond to changes in the environment as detected by the sensors.
- **Power supply:** The power supply provides the robot with the electricity it needs to operate.
- **Communication Interfaces:** Robots may need to communicate with external devices or systems. Communication interfaces such as Wi-Fi, Bluetooth, USB ports, and serial connections enable data exchange between the robot and other devices.
- **End Effectors:** If the robot is designed to interact with objects or perform tasks, it may have specialized end effectors or grippers. These are devices attached to the robot's manipulator that allow it to grasp, manipulate, or interact with objects.

2.5.2 Types of Robots

The robots may be classified based on where they are employed.

- **Industrial robots:** These robots are used in factories to perform repetitive tasks, such as welding, painting, and assembly. They are typically large and strong, and they can be programmed to perform a variety of tasks.
- **Service robots:** These types of robots are used to perform tasks that are helpful to humans, such as cleaning, delivering food, and providing customer service. They are typically smaller and more agile than industrial robots, and they can be programmed to interact with humans in a safe and efficient way.
- **Medical robots:** These robots are used in surgery and other medical procedures. They can be used to perform delicate tasks with greater precision than human surgeons.
- **Military robots:** These robots are used in warfare to perform activities that are considered dangerous or difficult for humans, such as bomb disposal and surveillance.
- **Research robots:** These robots are used to conduct research in a variety of fields, such as robotics, artificial intelligence and materials science.
- **Educational robots:** These robots are used to teach people about robotics and STEM subjects. They are typically small and easy to use, and they can be programmed to perform a variety of tasks.
- **Entertainment robots:** These robots are used for entertainment purposes, such as dancing, performing and companionship. They are typically designed to be interactive and engaging.
- **Manufacturing:** Robots are used in manufacturing to perform tasks such as welding, painting, and assembly. They can help to improve efficiency and accuracy, and they can also reduce the risk of injuries to workers.
- **Healthcare:** Robots are used in healthcare to perform tasks such as surgery, rehabilitation and patient care. They can help to improve the quality of care and reduce the cost of healthcare.
- **Customer service:** Robots are used in customer service to answer questions, provide support, and complete transactions. They can help to improve customer satisfaction and reduce the workload of human employees.
- **Military:** Robots are used in the military to perform tasks such as surveillance, bomb disposal and combat. They can help to protect soldiers and civilians, and they can also increase the effectiveness of military operations.
- **Space exploration:** Robots are used to explore space and perform tasks that are too dangerous or difficult for humans. They can help to gather data, conduct experiments and build structures on other planets.

2.5.3 Sophia

Sophia is a humanoid robot (Fig. 2.19) developed by Hong Kong-based Hanson Robotics. It gained international attention due to its lifelike appearance and advanced artificial intelligence capabilities. Sophia was designed to visually resemble a human and can simulate facial expressions and hold conversations with people.

Figure 2.19 Sophia is the first robot citizen

Key features of Sophia include:

- **Appearance:** Sophia is designed to mimic human features, with a face that can move its eyes, lips, and eyebrows, creating a range of facial expressions.
- **AI Interaction:** Sophia's artificial intelligence system enables it to engage in conversations with humans. It uses natural language processing to understand and respond to questions and statements.
- **Voice Recognition:** Sophia has the ability to recognize and process speech, allowing it to respond to verbal commands and questions.
- **Learning Abilities:** While not a fully autonomous learning machine, Sophia can learn and adapt to some extent through interactions. It can remember individual users and previous conversations.
- **Public Appearances:** Sophia has been showcased at various international events, conferences, and media appearances. It has even been granted citizenship in some countries, although this symbolic gesture has generated ethical discussions about robot rights and responsibilities.
- **Expression of Emotion:** Sophia can display a limited range of emotional expressions using its facial features, allowing it to convey happiness, surprise, sadness and other emotions.

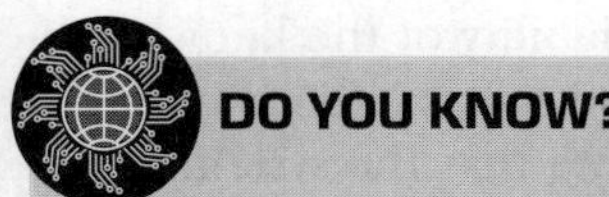

ImageNet is a large-scale image database widely recognized for its role in advancing computer vision and machine learning. Founded by Fei-Fei Li in 2007, ImageNet contains millions of labelled images in various categories. It gained fame through the ImageNet Large Scale Visual Recognition Challenge (ILSVRC), an annual competition that spurred the development of deep learning algorithms, particularly convolutional neural networks (CNNs). In 2012, the breakthrough came when a CNN-based approach achieved a significant drop in classification error. This marked a pivotal moment in the history of AI, paving the way for improved object recognition, image classification, and image analysis. ImageNet remains a valuable resource for researchers and has significantly contributed to the progress of computer vision technology.

2.6 KNOWLEDGE ENGINEERING

Knowledge engineering is a discipline within artificial intelligence (AI) and computer science that focuses on designing, building, and maintaining knowledge-based systems. The goal of knowledge engineering is to capture, represent, organize and utilize human expertise and domain-specific knowledge in a format that can be processed and manipulated by computers. This knowledge is then used to create intelligent systems that can reason, make decisions, and solve complex problems in specific domains.

2.6.1 Steps in Knowledge Engineering

1. **Knowledge Acquisition:** This involves gathering knowledge from domain experts and converting it into a format that can be understood by computers. Different techniques, such as interviews, surveys and analysis of existing documents, are used to capture the expertise of human experts.

2. **Knowledge Representation:** Once knowledge is acquired, it needs to be organized and structured in a way that can be easily interpreted by machines. Various formalisms, such as ontologies, semantic networks, rules, frames, and more, are used to represent knowledge in a format suitable for AI systems.

3. **Knowledge Integration:** Often, knowledge engineering involves integrating various sources of knowledge to create a comprehensive representation. This can include combining structured data, unstructured text and expert rules to build a cohesive knowledge base.

4. **Inference and Reasoning:** Knowledge engineering involves developing mechanisms and algorithms for drawing conclusions, making inferences, and reasoning with encoded knowledge. This allows the AI system to answer questions, solve problems and simulate human-like decision-making.

5. **Maintenance and Refinement:** As domains evolve, knowledge-based systems need to be updated and refined to remain accurate and relevant. Knowledge engineers are responsible for maintaining and improving the knowledge bases over time.

6. **Verification and Validation:** Ensuring the correctness and reliability of the knowledge base is crucial. Verification involves checking that the knowledge has been accurately captured and represented, while validation involves confirming that the system's outputs align with expected outcomes.

7. **Deployment and Interaction:** Knowledge engineering includes integrating the knowledge-based system into real-world applications and enabling interaction between the system and users. This often involves creating user interfaces that allow users to query the system and receive meaningful responses.

Among the above steps, while some are engineering steps, some steps need more discussion like Knowledge Representation.

2.6.2 Knowledge Representation

There are several different knowledge representation schemes used in artificial intelligence (AI) to capture and represent information and knowledge in a format that computers can understand and manipulate. Each scheme has its own strengths and weaknesses, making them suitable for different types of problems and domains. Here are some common knowledge representation schemes.

Semantic Networks

Semantic networks (Fig. 2.20) represent knowledge using nodes (concepts) and links (relationships) between them. This graphical representation is particularly useful for illustrating relationships and hierarchies within a domain.

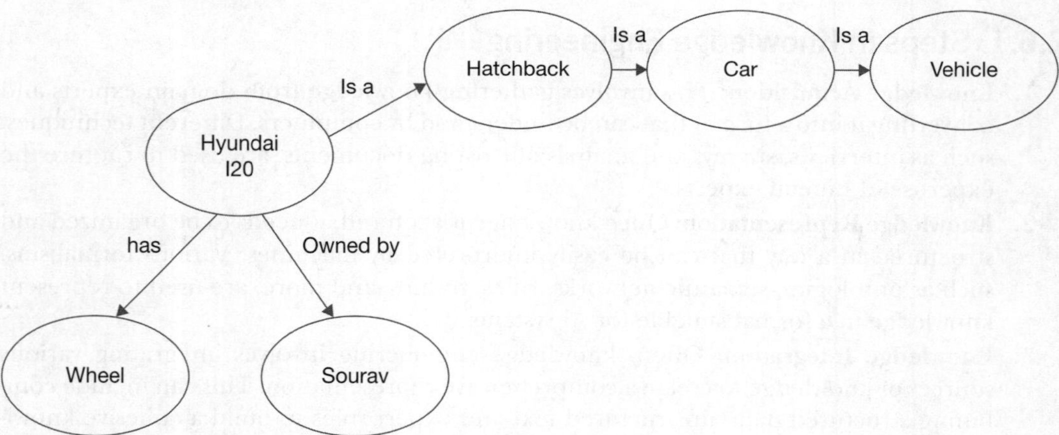

Figure 2.20 A sample semantic network

In the above diagram, a simple concept is presented. Hyundai i20 is a hatchback car, which is also car and also a vehicle. It has wheels and is owned by Sourav. While Semantic Network is a graphical representation with associated benefits, this lacks formalism. There is no clear guideline on types of relationships.

Frames

Frames organize knowledge into structured units called frames, which contain attributes, slots and values that describe an object or concept. Frames are useful for representing structured information and capturing inheritance relationships. If we take the same example, the car will be called a frame and different attributes of the frame like the make, model, year and colour will be called slots.

1. **Make:** Hyundai
2. **Model:** i20
3. **Year:** 2022
4. **Colour:** Red

For representing the concept of hierarchy, concepts like super frames, subframes may be used. For example, here car can be a superframe.

First-Order Logic (FOL)

FOL extends propositional logic by introducing variables, quantifiers and predicates. It provides a formal and expressive representation for relationships, rules and facts. Let's look at the components

- **Variables:** Variables are used to represent objects in the world. They are denoted by lower case letters, such as x, y, and z.
- **Constants:** Constants are used to represent specific objects in the world. They are denoted by upper case letters, such as A, B, and C.
- **Predicates:** Predicates are used to represent relationships between objects. They are denoted by capital letters, such as P, Q, and R.
- **Connectives:** Connectives are used to combine statements together. They include logical operators such as $\land$ (and), $\lor$ (or), $\rightarrow$ (implies), and $\leftrightarrow$ (iff).

FOL statements are constructed by combining these symbols using the following rules:

- A predicate can be applied to a variable or a constant to form a statement. For example, the statement $P(x)$ means that the object denoted by x has the property denoted by P.
- Statements can be combined using the connectives. For example, the statement $P(x) \land Q(y)$ means that the object denoted by x has the property denoted by P and the object denoted by y has the property denoted by Q.

Examples of Objects and Predicates:

- **Predicate:** L (x, y) meaning 'x Likes to eat y'
- **Statement:** 'Soham loves Pizza.'
- **FOL Representation:** L (Soham, Pizza)

Statement with Quantifiers:

- **Universal Quantification:** 'All men are Mortal.'
  - **FOL Representation:** $\forall x\ (Man(x) \rightarrow Mortal(x))$

- Where ∀x means 'for all x,' Man(x) represents 'x is a Man,' and Mortal(x) represents 'x is a Mortal.'
  - The arrow indicates implies
- **Existential Quantification:** 'There exists a red car.'
  - **FOL Representation:** ∃x (Car(x) ∧ Red(x))
  - Where ∃x means 'there exists an x,' Car(x) represents 'x is a car,' and Red(x) represents 'x is red.'

There are many other notations, like ¬Sunny(x), which stands for negation. Essentially, this means x, which represents that the day x is not sunny.

Rules/Production Rules

Rule-based systems use a set of if-then rules to represent knowledge. Production rules, also known as production systems, are a fundamental concept in the field of artificial intelligence (AI) and expert systems. They are used for representing knowledge and making decisions based on a set of condition-action pairs. Production rules consist of two main components:

Conditions

These are logical statements or tests that describe the circumstances or criteria that must be met for the rule to be applied. Conditions are typically expressed in terms of facts, data, or variables. They define when a particular rule should be triggered.

Actions

These are statements or instructions that specify what should be done when the conditions associated with a rule are satisfied. Actions can include making conclusions, updating facts, executing procedures, or triggering other rules.

Production rules are often written in the following format:

IF (Conditions)

THEN (Actions)

Here are a few examples of production rules:

Medical Diagnosis:

IF patient has a fever AND patient has a sore throat

THEN diagnose the patient with the flu AND prescribe antiviral medication

Inventory Management:

IF inventory level is below a certain threshold

THEN reorder the product in a specified quantity

Production rules are a powerful knowledge representation tool because they allow AI systems to encode domain-specific knowledge and expertise in a structured and understandable manner. Expert systems, which are AI systems designed to solve specific problems or make decisions in specialized domains, often use production rules to model the decision-making process. When applied sequentially or in parallel, production rules enable expert systems to reason, infer and provide solutions based on the available data and knowledge.

Few other important topics as far as knowledge engineering is concerned is Fuzzy logic.

2.6.3 Fuzzy Sets and Logic

A fuzzy set is a concept in mathematics and set theory that extends the idea of a crisp set. To understand a fuzzy set, it's helpful to first define a crisp set.

A crisp set is a well-defined set-in traditional set theory. It contains elements that either completely belong to the set or completely do not belong to the set. For example, in a crisp set of 'Even Numbers,' 2, 4, and 6 are fully part of the set, while 3, 5, and 7 are completely outside the set.

A fuzzy set is a more flexible and nuanced concept. In contrast to crisp sets, fuzzy sets allow elements to have degrees of membership. This means that elements can partially belong to the set to varying degrees. Instead of just 'in' or 'out' of the set, you can say an element is, member of a set with membership value of 0.7, indicating a high degree of membership.

Example

Let's consider a fuzzy set of 'Tall People' based on height. In a crisp set, you would say someone is either 'in' the set (tall) or 'out' of the set (not tall). However, in a fuzzy set, you can express degrees of tallness. For instance, a person who is 6 feet tall might have a membership degree of 0.8, indicating a high degree of being 'tall,' while someone who is 5 feet tall might have a membership degree of 0.3, showing a lower degree of 'tallness.' Fuzzy logic is a type of logic that uses fuzzy sets to represent knowledge and to make decisions. Fuzzy logic is based on the idea that there are many different degrees of truth, rather than just true and false.

Fuzzy logic is used for diverse applications, such as:

1. **Robotics:** Fuzzy logic is used to control robots in complex and uncertain environments.
2. **Machine learning:** Fuzzy logic is particularly useful for machine learning algorithms which have to deal with data that is noisy or incomplete.
3. **Natural language processing:** Fuzzy logic is used to construct natural language processing systems that can understand and generate human language in a more natural way.

Here is an example of how fuzzy logic can be used to make a decision:
Suppose you are designing a robot that needs to be able to navigate in a crowded environment. You could use fuzzy logic to represent the robot's knowledge about the environment, such as the distance to other objects and the speed of other objects. You could then use fuzzy logic to make decisions about how the robot should move. For example, the robot could use fuzzy logic to decide whether to turn left, turn right, or go straight.

Fuzzy logic is a very effective tool that can be used to model and reason about uncertainty. It is used in a variety of applications to make decisions in complex and uncertain environments.

DO YOU KNOW

Fuzzy String matching is a popular concept, where when a string is searched, instead of exact match (crisp), similarity score is calculated and similar strings are returned. This is called as partial matching. Suppose you are searching for the word 'cat' on a website. The website's search engine uses fuzzy string matching to find all of the pages on the website that contain words that are similar to 'cat', such as 'cats', 'kitten', and 'kitty cat'.

Rough set theory is a mathematical framework used to analyse data when you have incomplete or uncertain information. It helps you make decisions and classify data into different categories, even when you're not entirely sure.

Let's look at another important concept named as Rough Set.

Example 1: Imagine you have a collection of fruits, and you want to classify them as either 'Apples' or 'Oranges' based on their characteristics. You have information about their colour, taste and shape.

1. **Definitely Apples:** These are the fruits you're quite certain are apples based on the information you have. For instance, if a fruit is red, sweet, and round, you can confidently say it's an apple.
2. **Possibly Apples:** These are the fruits that you're not entirely sure about. They might be apples, or they might be oranges. For instance, a fruit could be red and round (like an apple), but you're not sure about its taste. It might be sweet (like an apple) or sour (like an orange).

In rough set theory, 'Definitely Apples' represents the lower approximation, and 'Possibly Apples' represents the upper approximation. The upper approximation includes fruits that might belong to the 'Definitely Apples' group, but you can't be completely certain.

Example 2: Now, let's say you have a dataset of cars, and you want to determine if they are 'Fuel-Efficient' or 'Not Fuel-Efficient.' You have information about their fuel efficiency (miles per gallon), engine size, and weight.

1. **Definitely Fuel-Efficient Cars:** These are the cars you're quite certain are fuel-efficient based on the information you have. For example, if a car has high fuel efficiency, a small engine, and is lightweight, you can confidently classify it as fuel-efficient.
2. **Possibly Fuel-Efficient Cars:** These are the cars that you're not entirely sure about. They might be fuel-efficient, or they might not be. For instance, a car might have high fuel efficiency (like a fuel-efficient car) but a large engine (which is not typical for fuel-efficient cars). It falls into the 'Possibly Fuel-Efficient' category because you can't be completely sure.

Rough set theory helps you work with data like this, where you have uncertainty, and it allows you to make decisions and classifications even when some information is incomplete or ambiguous.

The following table encloses the difference between fuzzy and rough sets:

Characteristic	Fuzzy set	Rough set
Definition	A set where the elements have degrees of membership between 0 and 1.	A set where there are some members that belong to the set with certainty and some members that may or may not belong to the set.
Uncertainty	Uses degrees of membership to represent the uncertainty in whether or not an element belongs to a set.	Uses crisp sets to represent the uncertainty of whether or not an element belongs to a set.
Knowledge representation	Typically used to represent knowledge about the world.	Typically used to extract knowledge from data.
Example	The fuzzy set of 'tall people'.	The rough set of 'customers who are likely to make a purchase'.

2.7 SUMMARY

- The ultimate goal of AI should be creating an artificial man/ robot.

- The subsystems for the same will be computer vision, natural language processing, robotics, machine learning and knowledge representation.

- The objective of computer vision is to enable computers to understand a natural scene like a human being can. Some important areas of computer vision, are object recognition, face recognition, medical imaging, etc.

- Similarly, the goal of NLP is to develop the ability of computers to understand language and generate responses in natural language. Some of the very critical tasks are text summarization, question answering, sentiment analysis etc.

- Machine learning concerns about the efficient algorithm for learning. The basic types are supervised learning where you have past examples and unsupervised learning which does not have past examples. Some important algorithms are k-means, k-NN, decision tree and logistic regression.

- One other sub-field of machine learning is called deep learning, whose basic unit is an artificial neuron, this is mimicked after a biological neuron.

- Deep learning debuted with limited success but then picked up as more data became available, there is significant algorithmic improvement and GPU.

- Reinforcement learning is a subfield, which is best suited for trial-and-error environment. The best policy is determined as the agent interacts with the environment through actions (For desired actions, agents receive reward and for not-so-desired actions the agent receives penalty), some of the important paradigms of reinforcement learning are multi-armed bandit, Q-Learning, etc.

- Some of the important components of robots are sensors, actuators, frames, communication interface, power supply, etc.

- Some of the important application domains where robots are being used are education, military, health, industrial automation, etc.

- Knowledge representation/ engineering is another important component. For representation techniques like Semantic Net, Frames can be used. Predicate and first-order logic can be used for reasoning. For dealing with uncertainty concepts like fuzzy set and logic and rough set can be employed.

2.8 PRACTICE EXERCISES

2.8.1 Subjective Questions (10 Marks)

1. What is the fundamental difference between intelligent thinking (cognitive intelligence) and intelligent action (behavioural intelligence) in the context of AI? Provide examples for each.

2. What are some of the most important capabilities that AI systems must have that function at the human or superhuman level? Give a thorough explanation of each capacity.

3. How do the different subfields of AI intersect and collaborate with each other to advance the field of artificial intelligence? Give examples of how advancements in one subfield can benefit others

4. Explain the concept of object recognition in computer vision. Provide examples of real-world applications where object recognition is used.

5. Describe the difference between face detection and face recognition in computer vision. Provide real-world examples of each of these techniques in detail.

6. Discuss the importance of scene understanding in computer vision and provide examples of how it is used in applications like self-driving cars and augmented reality.

7. How is computer vision technology being utilized in the field of medical imaging? What benefits and challenges does it bring to the healthcare industry?

8. Explore the potential applications and challenges of computer vision in the development of the metaverse and its impact on social interactions and digital experiences.

9. Explain the importance of text pre-processing in Natural Language Processing (NLP). Provide examples of common text pre-processing techniques.

10. Explain the key differences between biological neurons and artificial neurons in the context of deep learning. How do artificial neurons simulate the decision-making process of biological neurons?

11. What is the role of the universal approximation theorem in deep learning, and how does it relate to neural network architecture? Provide an example to illustrate its significance.

12. Describe the main steps in knowledge engineering and explain the importance of each step in the development of knowledge-based systems. Provide an example to illustrate the process.

13. Discuss the various knowledge representation schemes used in artificial intelligence, including Semantic Networks, Frames, First-Order Logic (FOL), and Production Rules. Compare and contrast their strengths and weaknesses, and provide a real-world example where each representation scheme might be suitable.

14. Explain the concepts of Fuzzy Sets and Logic in the context of knowledge engineering. How are fuzzy sets used to represent and process uncertain or vague information? Provide a practical application of fuzzy logic in a specific domain.

2.8.2 Subjective Questions (5 Marks)

1. What a multi-armed bandit problem is different from Q learning.

2. Explain the fundamental concepts of the multi-armed bandit problem in reinforcement learning.

3. Describe the concept of transfer learning in deep learning. How does it enable knowledge transfer between different tasks, and why is it more feasible in deep learning compared to traditional machine learning

4. Describe the concept of dimensionality reduction in machine learning. Why is it important, and what techniques are commonly used for this purpose?

5. Provide an overview of anomaly detection in machine learning. How can it be applied to real-world scenarios, and what are the key use cases?

6. Discuss the challenges and applications of Named Entity Recognition (NER) in NLP. Provide examples of how NER can be applied to real-world scenarios.

7. Explain the fundamental difference between supervised learning and unsupervised learning in machine learning. Provide examples for each.

8. What are the challenges associated with solving the "Long Tail Problem" in recommender systems? How can these challenges be addressed?

9. Explain how object recognition is used in Medical Imaging.

2.8.3 Short Answer-type Questions (2 Marks)

1. Describe the difference between face detection and face recognition in computer vision.

2. Enlist two key differences between rough set and crisp set.

3. Describe the difference between extractive and abstractive text summarization methods in NLP.

4. Differentiate between Intelligent Thinking with Intelligent Action.

5. Explain one important application of Computer Vision adopted in India.

6. Explain briefly how rule-based machine translation works.

7. What is collaborative filtering in context to recommender systems?

8. What do you understand by cold start problem?

9. What is the role of threshold in a neural network?

10. What is the specialization of CNN?

11. What is value function in Reinforcement Learning?

12. What do you understand by exploitation exploration trade off?

13. How Thomson Sampling differs from UCB?

14. What are the common sensors used in Robotics?

15. What are slots in context to frames?

2.8.4 Multiple-choice Questions (1 Mark)

1. What is the key difference between intelligent thinking and intelligent action in AI?
   (a) Intelligent thinking involves problem-solving, while intelligent action pertains to understanding natural language.
   (b) Intelligent thinking refers to interacting with the physical world, while intelligent action involves reasoning.
   (c) Intelligent thinking relates to simulating human thought processes, and intelligent action deals with manipulating the physical world.
   (d) There is no difference; both terms are used interchangeably in AI.

2. Which subfield of AI is responsible for processing visual cues and understanding images and videos?
   (a) Machine Learning
   (b) Knowledge Engineering
   (c) Natural Language Processing
   (d) Computer Vision

3. When AI systems are capable of self-learning and deriving rules from examples, which subfield does this represent?
   (a) Natural Language Processing
   (b) Knowledge Engineering
   (c) Machine Learning
   (d) Robotics

4. What does the subfield of robotics primarily focus on in AI?
   (a) Developing algorithms for data analysis
   (b) Interaction between computers and human languages
   (c) Creating intelligent robots that can interact with the physical world
   (d) Knowledge representation and logical inference

5. What is the primary goal of object recognition in computer vision?
   (a) To capture images of objects
   (b) To identify the presence of a human face
   (c) To gain an understanding of what is in an image
   (d) To create a realistic virtual environment

6. What is the key difference between face detection and face recognition in computer vision?
   (a) Face detection identifies the identity of a person based on their face, while face recognition focuses on the presence of a human face in an image or video.
   (b) Face detection is more advanced and specific, while face recognition is more basic and general.
   (c) Face detection is used in self-driving cars, while face recognition is used in augmented reality.
   (d) Face detection is used for identifying objects in the environment, while face recognition is used for medical imaging.

7. Which emerging concept is closely related to augmented reality and virtual reality and involves social interactions in a digital universe?
   (a) Self-driving cars
   (b) Object recognition
   (c) The metaverse
   (d) Scene understanding

8. Which of the following is not a common text pre-processing technique in NLP?
   (a) Lowercasing
   (b) Tokenization
   (c) Image segmentation
   (d) Punctuation removal

9. What is the primary goal of text classification in NLP?
   (a) Translate text from one language to another
   (b) Categorize text into predefined labels
   (c) Generate summaries of text
   (d) Identify and extract named entities from the text

10. Which machine learning task is focused on predicting continuous numerical values based on input data?
    (a) Classification
    (b) Clustering
    (c) Regression
    (d) Dimensionality reduction

11. Anomaly detection in machine learning is used for:
    (a) Categorizing data into predefined classes.
    (b) Predicting continuous values.
    (c) Identifying unusual or abnormal data points.
    (d) Reducing the dimensionality of data.

12. The universal approximation theorem in deep learning states that:
    (a) A single-layer neural network can approximate any function.
    (b) A neural network with multiple hidden layers can approximate any continuous function.
    (c) The choice of activation function is irrelevant in neural networks.
    (d) Neural networks can only approximate simple linear functions.

13. Transfer learning is more feasible in deep learning than in traditional machine learning because:
    (a) Deep learning models have fewer parameters.
    (b) Deep learning models require larger amounts of labelled data.
    (c) Deep learning models have more complex architectures.
    (d) Deep learning models can leverage existing knowledge for new tasks with little adjustments.

14. Which of the following strategies in the multi-armed bandit problem involves playing the machine with the highest estimated reward most of the time but occasionally exploring other machines?
    (a) Greedy strategy
    (b) Epsilon-greedy strategy
    (c) UCB (Upper Confidence Bound) strategy
    (d) Thompson sampling strategy

15. What is the primary goal of reinforcement learning?
    (a) To classify data into predefined categories
    (b) To minimize the difference between predicted values and actual outcomes
    (c) To learn how to behave in an environment by trial and error to maximize expected rewards
    (d) To generate realistic data using generative adversarial networks

16. What is the primary goal of knowledge engineering?
    (a) To design computer hardware
    (b) To develop algorithms for natural language processing
    (c) To capture, represent, and utilize human expertise and domain-specific knowledge for intelligent decision-making
    (d) To create artificial neural networks

2.9 ANSWER KEYS

2.9.1 Multiple-choice Questions

1. (c)	**2.** (d)	**3.** (c)	**4.** (c)	**5.** (c)	**6.** (a)	**7.** (c)	**8.** (c)
9. (b)	**10.** (c)	**11.** (c)	**12.** (b)	**13.** (d)	**14.** (b)	**15.** (c)	**16.** (c)

Industrial Applications of AI

In the last two chapters, we have been introduced to the basic concepts of AI, the different constituents (or sub-areas) that it has and also how the concepts have evolved over the past few decades. We have also briefly touched upon the industrial applications of AI, highlighting a few quick points of enablement in these areas. However, to appreciate the fact that AI has a significant impact in solving complex problems in different industries, a deep dive into some of the industries is needed. Only then, looking at a few salient use cases, we can appreciate the fact that AI is absolutely essential to solve a few of these problems. In this chapter, we will delve into the applications of AI in industries and see how this technology is not just a futuristic concept but a present reality that is fundamentally changing the landscape of commerce, healthcare, finance, manufacturing and beyond.

Across diverse industry sectors, some of which are shown in Fig. 3.1, AI applications are reshaping the way businesses operate, solve problems, and interact with their stakeholders. By combining computing capabilities with AI algorithms, we are witnessing a new era of efficiency, innovation, and adaptability.

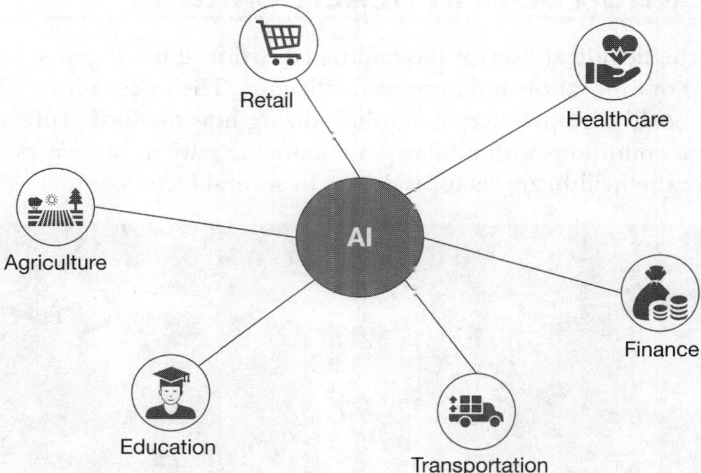

Figure 3.1 Applications of AI in Industry

In the realm of e-commerce specifically, AI has become a tool for enhancing customer experience by streamlining operations and predicting market trends. E-commerce platforms utilize AI algorithms to analyse consumer behaviour patterns and anticipate preferences to offer recommendations. Virtual assistants equipped with natural language processing capabilities provide customer interactions by handling queries and facilitating purchases. Moreover, AI-powered chatbots play a role in customer service by promptly addressing inquiries and ultimately improving overall satisfaction. The integration of AI into commerce not only optimizes business processes, it also creates an intuitive and responsive environment for consumers.

The healthcare industry is at the forefront of incorporating AI technology, which has brought about changes in diagnostics, treatment plans and patient care. AI algorithms are now able to analyse large amounts of medical images enabling fast diagnoses for conditions ranging from cancer to neurological disorders. In this sector, AI has become a game changer by providing insights, risk assessments and operational efficiency.

As we explore the applications of AI across industries, it becomes clear that its transformative power extends beyond specific sectors. The combination of data analytics, machine learning and intelligent automation is reshaping how businesses operate by challenging traditional models and emphasizing adaptability and innovation. In the following sections, we will delve deeper into examples and case studies that highlight the impact of AI in each industry while shedding light on what lies in this era of unprecedented technological advancement.

POINTS TO PONDER

Did you know that the International Data Corporation (IDC) released a report stating that the revenues, from sales of hardware, software and services in the field of AI are growing at a compound growth rate (CAGR) of 17.5%? It is predicted that this market will surpass $500 billion by 2024!

3.1 APPLICATION OF AI IN HEALTHCARE

In recent years, the healthcare sector has embraced Artificial Intelligence (AI) leading to an era characterized by innovation and improved efficiency. The integration of AI (as resembled in Fig. 3.2) into healthcare practices is revolutionizing how medical professionals diagnose, treat and manage conditions with a focus on enhancing patient outcomes. The widespread adoption of AI in the healthcare sector is driven by several factors.

Figure 3.2 Integration of AI in Healthcare

1. The healthcare industry generates an abundance of data including images, patient histories and prescriptions. This data intensive nature of healthcare industry makes it necessary to utilize automation to analyse and draw insights from this information.
2. Healthcare data has high variability complexity. It is also extremely intense a manual process to analyse and infer from medical data. So, automation is really needed to complement manual analysis.

This section delves into the multifaceted applications of AI in healthcare, highlighting its role in diagnostics, personalized medicine, administrative tasks and patient care.

3.1.1 Role of AI in Medical Diagnosis

One specific area where AI has made significant contributions is diagnosis. By employing machine learning algorithms, AI has transformed the way diagnoses are made in healthcare. These algorithms have proven their capabilities in analyzing datasets, interpreting images and identifying subtle patterns that aid traditional diagnostic methods. Following are some of the areas of applications of AI in medical diagnosis.

Radiology and Imaging

AI has made advancements in the field of radiology and medical imaging. By training machine learning models on large collections of images, AI algorithms have exhibited exceptional abilities to detect anomalies and identify patterns associated with various diseases. Overall, AI's role, in healthcare extends beyond diagnostics to include medicine, administrative tasks and patient care. In the case of breast cancer, for example, AI-powered analysis of mammograms has shown high sensitivity and specificity helping radiologists detect it early and accurately. Moreover, AI algorithms are utilized in the examination of X-rays, MRIs and CT scans (as resembled in Fig. 3.3) to not only speed up the process, but also enhance overall precision.

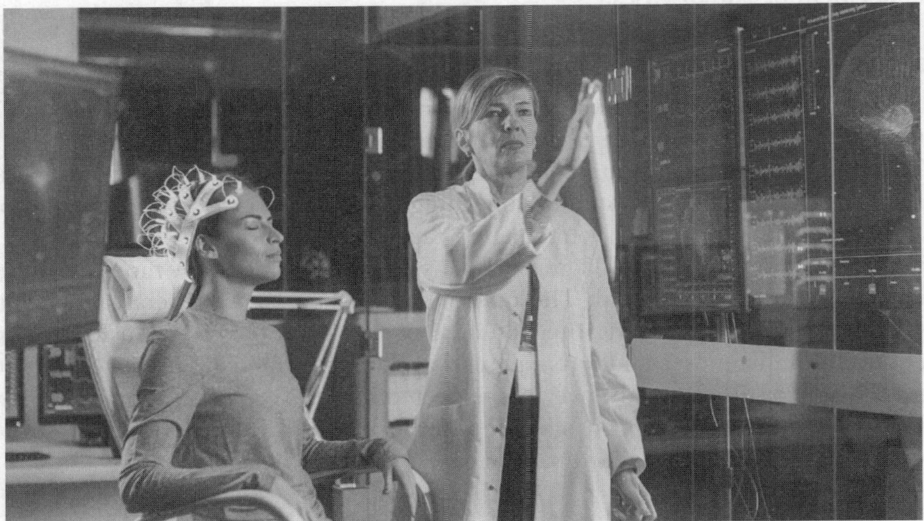

Figure 3.3 Role of AI in Medical Imaging

Retinal Imaging and Disease Detection

Moving on to retinal imaging and disease detection, AI can play a role in early identification of diseases like diabetic retinopathy and age-related macular degeneration (AMD). By analysing scans, AI algorithms can detect changes that indicate these conditions with a higher level of accuracy than traditional diagnostic methods. Glaucoma diagnosis and management also benefit from AI technology as it is one of the leading causes of blindness. In cataract surgery planning, AI is streamlining processes and improving outcomes by analysing data such as thickness and axial length. AI algorithms also predict the most suitable intraocular lens power for individual patients. This personalized approach enhances the accuracy of cataract surgery outcomes by reducing errors while increasing overall patient satisfaction.

Pathology and Histopathology

In pathology and histopathology, AI is transforming the way that tissue samples and pathology slides are analysed. Automated image recognition algorithms coupled with pattern detection, help pathologists identify abnormalities efficiently while ensuring diagnoses. AI-driven systems in the field of pathology have shown promise in detecting cancer cells, leading to a reduction in incorrect diagnosis and improved efficiency in pathology laboratories.

Dermatology and Skin Conditions

Moving on to dermatology, AI applications offer solutions for diagnosing skin conditions. By utilizing datasets of skin images, AI algorithms can accurately classify various dermatological conditions, including skin cancers. Teledermatology, enabled by AI, allows patients to remotely submit images of their skin lesions for analysis, enabling assessments and interventions. This not only improves accessibility to expertise, but also contributes to early detection and management of skin diseases.

Cardiology and ECG Analysis

In the area of cardiology, AI is making progress in analysing electrocardiograms (ECGs) for the detection of cardiovascular diseases. With the ability to identify patterns and deviations in ECG readings that may indicate conditions like arrhythmias or heart abnormalities, AI algorithms greatly aid accuracy. Additionally, integrating AI into cardiology enables patient monitoring, offering an approach to managing cardiovascular health.

The incorporation of AI into diagnosis opens up opportunities for more accurate, efficient and timely disease detection. The collaboration between AI systems and healthcare professionals is crucial in utilizing the potential of this technology. It can lead to patient outcomes accurate diagnoses and a reimagined approach to medical practices. As artificial intelligence (AI) continues to advance, it is expected to have a role in shaping the future of healthcare. One area where AI can make a difference is in the field of diagnosis enabling precision medicine and personalized patient care.

DID YOU KNOW?

- Back in 2015, Google's DeepMind collaborated with Moorfields Eye Hospital in London to make significant advancements in healthcare. Their AI algorithm showed remarkable accuracy in analyzing retinal scans and identifying different eye diseases. By 2017, their algorithm achieved a 94% accuracy in diagnosing eye conditions.
- In 2018 the U.S. Food and Drug Administration (FDA) approved the AI based system called IDx-DR developed by IDx Technologies. This system became authorized to make decisions without human intervention.
- The COVID 19 pandemic has emphasized the importance of utilizing AI in healthcare. AI-powered tools have played a role in diagnosing COVID 19 cases tracking its spread and predicting resource requirements. For example, scientists Mount Sinai Health System has created an AI algorithm that can accurately forecast the deterioration of COVID 19 patients up to 48 hours before it happens.

3.1.2 Role of AI in Early Detection and Disease Prevention

The use of machine learning algorithms and advanced data analytics empowers healthcare professionals to proactively anticipate, identify and mitigate health risks. AI plays a role in disease detection and intervention.

Early Disease Detection and Intervention

One of the primary applications of AI in predictive analytics is the early detection of diseases (refer to Fig. 3.4).

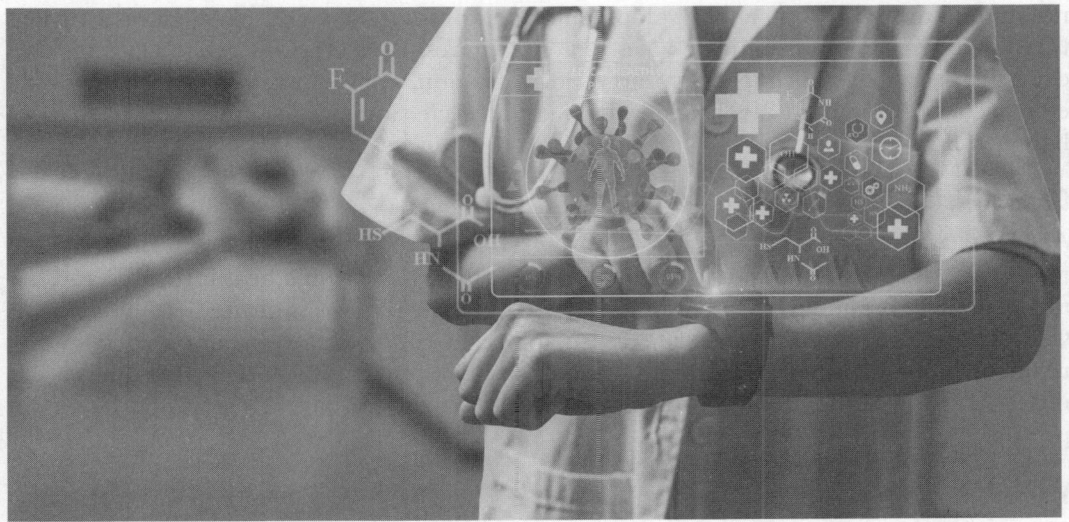

Figure 3.4 Role of AI in Early Detection of Diseases

By analysing datasets that include health records, genetic information and lifestyle data, AI algorithms can identify patterns that indicate the onset of diseases. For instance, predictive analytics driven by AI can assess an individual's risk based on factors like age, genetic pre-disposition and lifestyle choices. Early detection allows healthcare providers to intervene before symptoms appear and disease progression occurs ultimately improving outcomes.

Personalized Risk Assessment and Prevention Strategies

AI's advanced analytics capabilities enable the personalized assessment of risk for individuals. Considering markers such as history and lifestyle choices, AI algorithms can generate customized risk profiles. This valuable information empowers healthcare professionals to develop prevention strategies tailored to the needs of each individual. These strategies may include targeted interventions, lifestyle adjustments and screening protocols.

Moreover, by analysing data related to symptoms, medication adherence and lifestyle behaviours, AI can predict disease aggravations or complications. This foresight allows healthcare providers to take measures such as adjusting medications or providing lifestyle recommendations to effectively manage conditions. The result is a patient-centred approach to healthcare that prioritizes prevention than reactive treatment.

Disease Surveillance and Outbreak Prediction

Machine learning models can monitor data sources like travel patterns, climate conditions and social media activity to identify outbreaks of diseases. For example, during the COVID-19 pandemic, AI models were used to track the virus spread and predict areas at risk of outbreaks. This information helped allocate resources and mitigate the impact of diseases on populations through timely responses from public health authorities.

3.1.3 Role of AI in Drug Discovery and Development

The field of pharmaceuticals is undergoing a transformation with the integration of AI technology into drug discovery and development. By harnessing the power of AI-driven analytics, traditional and time-consuming processes are being reshaped. Various applications of AI in drug discovery have an impact on target identification, compound screening and overall effectiveness throughout the drug development lifecycle. Some of the areas have been highlighted below.

Clinical Trial Optimization

AI-driven analytics plays a role in optimizing clinical trials by enhancing patient recruitment, trial design and endpoint prediction. By analysing electronic health records and historical trial information, AI algorithms can identify participants and predict how patients will respond to treatments. This data-driven approach improves the efficiency of trials while reducing costs and timelines. Additionally, AI aids in identifying biomarkers that can serve as indicators for treatment efficacy enabling decision-making during the clinical development process.

Virtual Drug Design

The introduction of AI-driven analytics has brought about groundbreaking advancements in drug design. Virtual screening entails using models to analyse chemical databases and forecast the binding affinity between molecules and specific targets (as depicted in Fig. 3.5). By employing AI algorithms, potential drug candidates with desired properties can be identified, leading to time and resource savings compared to experimental screening. This approach facilitates the discovery of compounds and repurposing of existing drugs for therapeutic indications thereby maximizing the efficiency of drug development.

Figure 3.5 Role of AI in Drug Discovery and Development

Safety Assessment

AI contributes to predictive analytics for drug safety assessment by analysing large datasets to predict potential adverse effects and toxicity of drug candidates. By considering factors such as chemical structure, pharmacokinetics, and biological interactions, AI models can forecast potential safety issues early in the drug development process. This proactive approach enables researchers to prioritize safer compounds, reducing the likelihood of late-stage failures and ensuring that only the most promising candidates progress through the pipeline.

3.1.4 AI-powered Virtual Medical Assistant (VMA)

AI-powered virtual assistants are designed to enhance patient care, streamline processes, and provide valuable health-related information. AI empowers VMAs with the ability to understand natural language, interpret medical data, and deliver personalized insights. Following are some of the use cases related to AI-powered VMAs.

Patient Engagement and Communication

Virtual Medical Assistants powered by AI (Fig. 3.6), excel in engaging patients through their natural language processing capabilities. Patients can have interactions with these VMAs to seek information about symptoms, medications or treatment plans. The AI algorithms are designed to understand the context enabling context-aware conversations. This not only enhances engagement but also creates a more accessible and user-friendly healthcare experience.

Figure 3.6 AI-powered Virtual Medical Assistant

Medication Management and Adherence

AI-equipped VMAs play a role in managing medications by providing medication reminders information on drug interactions and personalized support for adherence. By analysing data, the AI algorithms predict adherence patterns and identify potential obstacles to medication compliance. Through timely interventions, VMAs minimize the risk of adverse events.

Appointment Scheduling and Administrative Tasks

The integration of AI into VMAs extends to streamlining tasks such as appointment scheduling and workflows. Patients can interact with VMAs to conveniently schedule appointments, receive reminders and access information seamlessly. This not only improves the efficiency of healthcare operations, but also enhances the overall patient experience by reducing administrative burdens.

3.1.5　AI-powered Robotics in Healthcare

The convergence of AI and robotics has ushered in an era in healthcare. The collaboration between AI and robotics has had an impact on the medical field. It has brought about advancements in practices leading to more precise surgeries, efficient diagnostics and personalized patient care. Below are some examples of how AI and robotics are being applied in healthcare.

Robotic Surgery and Precision Procedures

The integration of AI-driven robotics has revolutionized the field of surgery by offering levels of precision and capabilities. Surgical robots guided by AI algorithms assist surgeons in performing procedures with enhanced accuracy, as shown in Fig. 3.7. One notable example is the da Vinci Surgical System, which utilizes AI to translate surgeon movements into actions resulting in minimally invasive surgeries that lead to quicker recovery times. The use of AI algorithms also enhances real-time decision-making during surgeries in procedures like tumour removal and organ transplants.

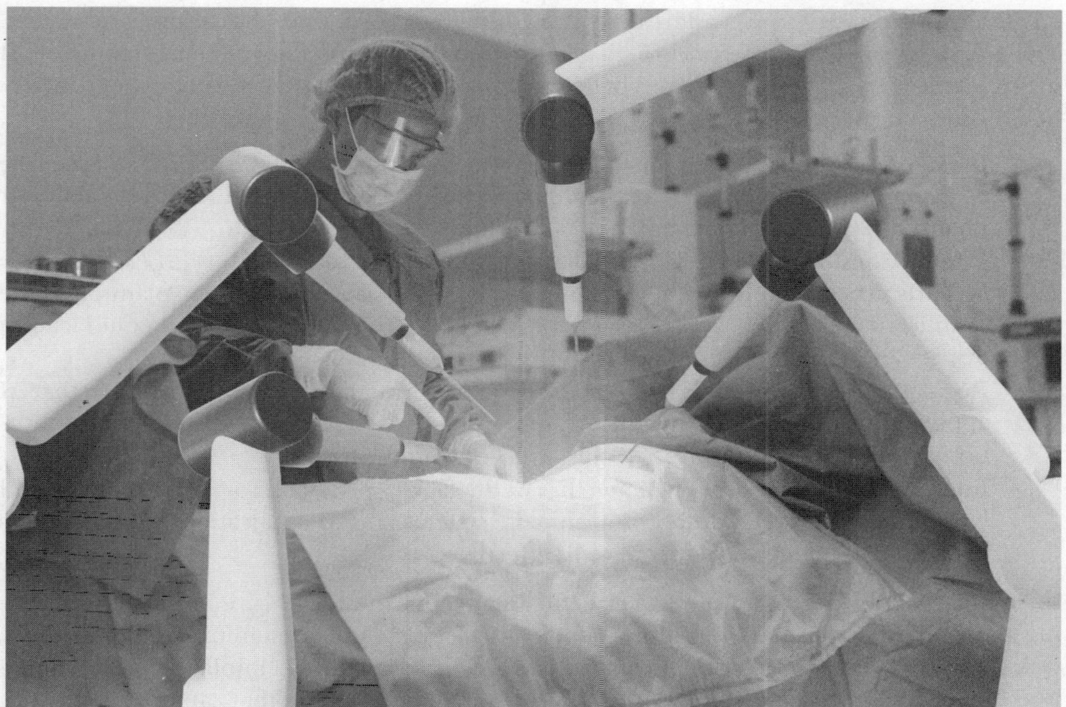

Figure 3.7 AI-powered Robotic Surgery [Doctors at Keck Medicine of USC perform the first robotic surgery to remove kidney cancer]

Rehabilitation Robotics and Physical Therapy

In the realm of rehabilitation, AI-powered robotics play a role in assisting and optimizing therapy for patients recovering from injuries or surgeries. Robotic exoskeletons and rehabilitation devices equipped with AI algorithms adapt to the needs of each individual, providing personalized rehabilitation programs. These robotic systems analyse patient movements, adjust resistance levels accordingly and offer real-time feedback to optimize the recovery process effectively. The application of AI-driven rehabilitation robotics has shown results for patients, especially with neurological disorders or mobility impairments.

Robotic-Assisted Personalized Care

Robotics, combined with AI technology, plays a role in providing patient care. By analysing health data, AI-powered robots can customize treatment plans to suit each patient's needs. These robots can administer medications, monitor vital signs and offer companionship in an adaptive manner. This tailored approach enhances patient comfort, adherence to treatment plans and overall well-being in long-term care situations.

3.1.6 Challenges of AI in Healthcare

While AI holds promise for healthcare applications, there are important concerns and challenges that need to be addressed when integrating AI into the healthcare system. The following are areas of concern that must be resolved for the implementation of AI in healthcare. By addressing these challenges, we can ensure patient wellness and better healthcare delivery.

- **Data Privacy and Security:** One major concern revolves around the privacy and security of patient data within AI-driven healthcare systems. The potential risk of breaches or unauthorized access to information is considered a primary threat.

- **Ethical Considerations:** Ensuring the use of AI in decision-making processes is particularly crucial when it comes to areas, like diagnosis and treatment planning. One significant area of focus involves addressing biases in AI algorithms that could lead to disparities in the delivery of healthcare services.

- **Regulatory Compliance:** It is crucial to navigate through frameworks to ensure that AI applications comply with healthcare standards. Adhering to data protection regulations, such as GDPR in Europe, adds layers of complexity.

- **Interoperability:** A major challenge lies in the lack of protocols for data exchange and collaboration between different healthcare systems and AI applications. This hampers the ability to integrate AI-driven tools effectively.

In summary, the integration of Artificial Intelligence, in the healthcare industry is reshaping the ecosystem. AI plays a role in enhancing accuracy and streamlining administrative tasks, ultimately leading to patient-centred care. With advancing technology, the collaboration between AI and healthcare presents the potential for solutions that can improve health benefits and pave the way for care and precision medicine. It is crucial for healthcare professionals, policymakers and the wider community to embrace and fully leverage AI's power to unlock its potential for promoting healthcare.

3.2 APPLICATION OF AI IN FINANCE

AI has also made an impact on the finance sector by revolutionizing financial operations. Its introduction of algorithms and advanced analytics has transformed how things work in this domain, some of them depicted in Fig. 3.8. The applications of AI in finance are vast offering solutions that streamline processes, enhance decision-making capabilities and improve customer experiences. One notable area where AI excels is algorithmic trading. Through models that leverage data and market trends, high speed trades can be executed efficiently leading to an increase in analysis driven strategies like high frequency trading.

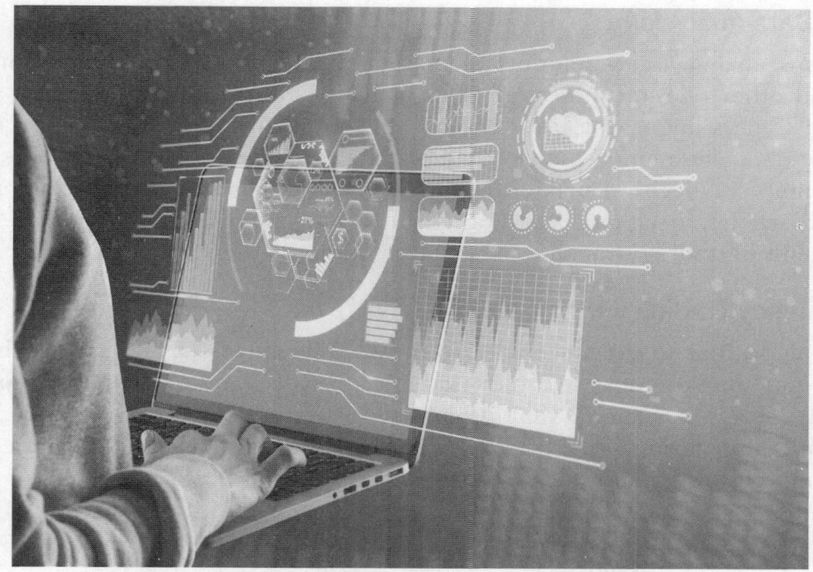

Figure 3.8 Application of AI in Finance

Furthermore, risk management has greatly benefited from AI implementations. Credit scoring and fraud detection have witnessed improvements with the assistance of AI models analysing datasets that include transaction history as well as social media activity. This enables real-time evaluation of creditworthiness while identifying any anomalies that may pose risks—providing institutions with robust defences against such risks. AI has greatly improved customer service. Virtual assistants and chatbots have completely transformed support systems by offering help and automated tasks. This ultimately enhances the experience, for customers. Moreover, AI-powered robo-advisors have made personalized banking and financial planning more accessible than before. These advisors provide customized investment advice and portfolio management based on preferences and financial goals.

AI is also playing a role in credit underwriting by automating decision-making processes. It incorporates data sources to provide a more comprehensive assessment of credit risk. AI applications have made regulatory compliance more efficient as well. The analysis of datasets using algorithms, ensures adherence to regulations benefiting processes like anti-money laundering (AML) and know-your-customer (KYC). The integration of AI in the finance sector also addresses concerns regarding data security. Advanced algorithms are leveraged for real-time detection and prevention of cyber threats. However, it is crucial to consider regulatory and privacy aspects.

3.2.1 AI in Algorithmic Trading

Algorithmic trading, also known as algo trading or automated trading, trading involves using computer algorithms to execute trading strategies with speed and efficiency. AI plays a role in this domain by providing analytical capabilities and automation features. Here are some key aspects related to AI-enabled trading.

Historical Data Analysis

AI algorithms analyse large amounts of market data such as price movements, trading volumes and other relevant indicators. This analysis helps to identify patterns and trends that human traders might not easily notice. Machine learning techniques are used by AI models to predict market movements based on data. By understanding these patterns AI can provide insights for decision making in trading. Additionally, AI-powered algorithms can analyse news articles, social media and other sources to gauge market sentiment. This allows for adjustments in trading strategies to capitalize on market sentiment.

High-Frequency Trading (HFT)

AI enables algorithms to execute trades (representative image in Fig. 3.9), such as buying and selling quantities of stocks and shares at high speeds - sometimes within microseconds or milliseconds. This is particularly vital in high frequency trading since swift order execution is crucial for taking advantage of price differences. AI algorithms optimize trade routing to reduce latency ensuring that trades are executed as swiftly as possible.

Figure 3.9 AI in Algorithmic Trading

Risk Management

AI algorithms continuously assess risk factors and adjust trading parameters to minimize potential losses. This includes monitoring market volatility, liquidity levels and other indicators of risk. With the help of AI-powered algorithms, stop loss orders can be automatically triggered when predefined risk thresholds are reached—thus reducing the impact of market movements.

Adaptive Trading Strategies

AI algorithms have the ability to adapt and evolve based on changing market dynamics. By continuously learning from new data, algorithms remain relevant and effective in various market conditions. AI also enables the simulation of trading strategies using data to evaluate their performance. This assists in refining strategies before implementing them in real market situations. It can learn from both successful and unsuccessful attempts, improving strategies over time to enhance performance.

In summary, AI enhances trading by utilizing data analysis, machine learning and automation. Its ability to process bulk amounts of data to adapt with changing market conditions and execute trades quickly, makes AI a powerful tool.

3.2.2 AI in Financial Risk Management

AI significantly enhances risk management by providing tools and capabilities for analysing, predicting and mitigating various types of risks. The application of AI in financial risk management is broad and covers different areas. Some key areas are described below.

Credit Scoring and Underwriting

AI models analyse datasets that include credit information as well as alternative sources like social media and transaction history to assess the creditworthiness of individuals and businesses. Based on these algorithms the credit underwriting process is automated, enabling real time decisions on loan approvals based on an evaluation of risk factors.

Fraud Detection

AI utilizes machine learning algorithms to identify patterns and anomalies in transactions aiding in the detection of potentially fraudulent activities. By analysing data and user behaviour, AI can establish patterns enabling the identification of deviations that may indicate fraudulent activities. AI-based fraud detection systems can operate continuously, monitoring transactions and user activities as they occur.

Compliance and Regulatory Risk

AI automates processes such as Anti Money Laundering (AML) and Know Your Customer (KYC) by analysing vast amounts of data ensuring adherence to regulatory requirements. Additionally, AI helps financial institutions stay aligned on changes by analysing and adapting policies to ensure compliance.

The benefits of adopting AI-powered risk management solutions are as follows:

- Faster detection and response to fraudulent transactions
- Enhanced accuracy and reduced False Positives
- Increased flexibility and agility through the ability to adapt to new and evolving risks
- Cost Savings due to minimized human intervention required

3.2.3 AI-based Customer Service

Chatbots

AI-powered chatbots and virtual assistants offer personalized customer support by addressing inquiries and handling tasks, thereby enhancing customer experience. AI has revolutionized customer interactions by using natural language processing (NLP) to automate communication. This makes it easier for customers to access information and carry out transactions.

Robo-advisors

Robo-advisors (representative image in Fig. 3.10) powered by AI offer personalized investment advice, portfolio management and financial planning services. These advisors analyze customer data and preferences to tailor their recommendations. By understanding customer behavior and data they can predict financial needs and provide tailored products and services.

Figure 3.10 AI in Customer Service

Personalized Financial Planning

AI considers individual goals, such as retirement planning, education funding, or saving for a home, to tailor investment strategies to specific objectives. Recommendations are adjusted based on life events like job changes, marriages, or the birth of a child to align with changing financial priorities.

When it comes to personalized financial planning, AI considers goals like retirement planning, education funding or saving for a home. Investment strategies are customized based on objectives. Additionally, recommendations are adjusted as life events such as job changes, marriages or the birth of a child occur, aligning with changing priorities.

3.2.4 Challenges in Application of AI in Finance

Implementing AI in the finance industry does come with challenges and risks. These range from ethical concerns to technical limitations. Some of the perceived challenges associated with the integration of AI in finance are presented below.

- **Handling Sensitive Information and Adversarial Attacks:** One major challenge is handling information, especially protecting them against adversarial attacks, since the finance industry deals with highly confidential personal and financial data. It is crucial for AI systems to have security measures in place to prevent data breaches or unauthorized access. AI models can be susceptible to attacks where malicious individuals manipulate input data to deceive the system and produce predictions or decisions.
- **Regulatory Compliance:** In the finance industry, there are regulations in place. However, keeping up with evolving frameworks and complying with existing regulations can be a challenge for AI applications. This is especially true as regulations may not have kept pace with advancements.
- **Overreliance on Historical Data:** One issue that arises when using AI models trained on data is their reliance on it. These models may struggle to adapt to shifts in market conditions or unforeseen events like crises or global pandemics. Financial markets are dynamic and ever-changing which can pose difficulties for models that were primarily trained on historical data.
- **Ethical Concerns:** Ethical concerns also come into play when using AI algorithms. These algorithms have the potential to inherit biases from the training data leading to discriminatory outcomes. It is crucial to make efforts to address bias within algorithms to ensure fair treatment across demographics. Additionally, lack of transparency in how AI models make decisions can undermine trust from users and hinder regulatory compliance. Therefore, it is important to understand and explain how these models arrive at decisions.
- **Complexity of Models and Interpretability:** Interpreting AI models, such as deep neural networks, can be complex and challenging due to their intricate nature. Lack of explainability can hinder comprehension, making it challenging to clarify model decisions to stakeholders including regulators and customers. Institutions are responsible for the outcomes of AI-based applications.
- **Operational Risks:** Mistakes in algorithms or unforeseen behaviour can result in financial losses. To minimize the risk of failures, it is essential to implement safe measures and conduct thorough testing procedures. Additionally, relying heavily on AI systems introduces risks like system outages or technical malfunctions that can disrupt services and operations.
- **Market Manipulation:** AI algorithms may be vulnerable to manipulation or exploitation by traders who aim to exploit the system for financial gain. This could potentially lead to market manipulation.

Addressing these challenges necessitates an approach that encompasses ethical considerations, ongoing monitoring, adaptation of regulatory collaboration, as well as a commitment to transparency and fairness in applying AI within the finance industry.

3.3 APPLICATION OF AI IN RETAIL

AI has transformed the retail industry on various fronts by providing solutions that enhance operational efficiency and improve customer experiences. Retailers utilize AI for demand forecasting and inventory management, leveraging analytics to optimize stock levels and dynamically adjust inventory. Personalization is a focus as AI-driven recommendation engines tailor product suggestions based on customer behaviour and preferences. AI has greatly enhanced search (Fig. 3.11) and try-on experiences. This means that customers can now search for products using images and even virtually try them on. Furthermore, AI has proven to be incredibly valuable, in optimizing supply chains. Through maintenance and route optimization businesses can ensure smooth and efficient operations.

Figure 3.11 AI in Retail Industry

Customer engagement is improved using AI-driven chatbots and virtual assistants which offer instant support for order tracking and personalized assistance. AI plays a role in enhancing security in physical retail environments. Retailers also utilize AI-powered dynamic pricing algorithms that adjust prices based on market conditions and demand. Analytics tools driven by AI analyse customer behaviour and provide insights for making informed decisions. Moreover, AI finds applications in smart store layouts and shelf management leading to a transformation of the industry. As technology continues to evolve, retailers are constantly exploring ways to integrate AI into their operations for growth and competitiveness.

3.3.1 Inventory and Store Layout Management

Dynamic Inventory Optimization

AI assists in dynamically adjusting inventory levels based on real-time data to prevent overstocking or stockouts (representative image in Fig. 3.12). Retailers rely on AI-powered algorithms that analyse data, seasonal trends and external factors to accurately predict demand.

Supply Chain Optimization

AI improves logistics and distribution processes by optimizing delivery routes for cost-effective and timely shipments. Furthermore, AI plays a role in enhancing the efficiency of supply chain operations by providing real-time insights and analytics for well-informed decision-making. These advancements not only streamline processes, they also contribute to building a resilient and agile supply chain enabling retailers to quickly adapt to market fluctuations and effectively meet customer expectations.

Figure 3.12 AI in Inventory and Store Management

Smart Store Layouts and Shelf Management

In terms of store management, AI-powered sensors and cameras analyse customer movement within the store to gain insights into areas. This helps to optimize store layouts for customer flow. Additionally, AI assists in managing inventory on shelves by ensuring that products are stocked according to demand and popularity.

3.3.2 Personalized Shopping Experience

Personalized Recommendation

To offer a shopping experience, AI-driven recommendation engines analyze customer behaviour purchase history and preferences. This analysis allows for product suggestions leading to increased customer engagement and sales. Furthermore, retailers can leverage AI to create targeted marketing campaigns by segmenting customers based on their preferences and behaviours.

Figure 3.13 AI Enabling Interactive Retail

Visual Search and Virtual Try-On

AI also brings value through search capabilities that enable customers to search for products using images. This enhances user experience and simplifies the product discovery process. Additionally, AI-powered augmented reality applications allow customers to virtually try on clothing items, accessories (refer to Fig. 3.13) or even makeup when shopping online. This feature enhances the online shopping experience while reducing return rates.

Customer Insights and Analytics

In the industry, AI tools play a role, in analysing customer behaviour both online and in-store. They provide insights into customer preferences buying patterns and emerging trends empowering retailers to make business decisions. By examining media, fashion blogs and other sources AI can even predict trends and help retailers stay ahead in the competitive market.

3.3.3 Customer Support

Chatbots and Virtual Assistants

When it comes to customer support AI AI-powered chatbots are highly effective. They offer efficient assistance by answering queries helping with product searches and handling routine tasks. This significantly improves customer satisfaction. Additionally, virtual assistants equipped with AI capabilities can provide real-time updates on order status, shipping details and delivery schedules.

Voice-Activated Shopping

Another innovative application of AI in retail is voice-activated shopping. By leveraging AI-driven voice assistants, customers can enjoy a shopping experience. They can effortlessly make purchases. Inquire about product details using language.

Fraud Protection

Fraud protection is another area where AI proves to be invaluable for retailers. Advanced algorithms analyse transaction patterns to detect any anomalies that may indicate activities. This does not safeguard the interests of customers. Also protects retailers from financial losses.

3.3.4 Challenges in Application of AI in Retail

While the application of AI, in the industry offers benefits mentioned above it does come with its own set of challenges that need to be addressed. To effectively tackle these challenges, a well-thought-out and strategic approach is necessary. It involves a combination of technology implementation, adherence to requirements, ethical considerations and organizational preparedness. As the retail industry continues to embrace AI, it becomes crucial to overcome these obstacles in order to fully harness the benefits of AI applications. Here are some of the challenges that need attention:

- **Privacy Concerns:** The use of AI in retail, particularly in customer-centric applications, raises privacy concerns. Customers may be hesitant to share personal information, and retailers need to navigate the delicate balance between personalization and respecting privacy.
- **Customer Acceptance:** Customers may have reservations about AI-driven technologies, especially when it comes to tasks traditionally handled by humans, such as customer service. Building trust and ensuring a positive user experience are essential for widespread customer acceptance.
- **Regulatory Compliance:** The evolving landscape of data protection and privacy regulations poses challenges for retailers using AI. Staying compliant with regulations such as GDPR and ensuring responsible AI practices are crucial but can be complex.

The integration of AI in retail is a dynamic process, with continuous advancements contributing to improved efficiency, enhanced customer experiences, and increased competitiveness in the industry. As technology evolves, retailers are likely to explore even more innovative ways to leverage AI for sustainable growth and operational excellence.

3.4 APPLICATION OF AI IN AGRICULTURE

The application of Artificial Intelligence (AI) in agriculture (refer to Fig. 3.14) has ushered in a new era of innovation, fundamentally transforming traditional farming practices and addressing long-standing challenges. Through the integration of advanced technologies such as machine learning, computer vision, and the Internet of Things (IoT), AI has facilitated the emergence of precision agriculture (refer to Fig. 3.15), enabling farmers to make data-driven decisions for optimal crop management. One of the primary applications lies in precision farming, where AI algorithms process real-time data from sensors, satellites, and drones, offering insights into soil conditions, crop health, and weather patterns. This empowers farmers to implement targeted strategies for irrigation, fertilization, and pest control, ultimately enhancing resource efficiency and crop yields.

Figure 3.14 AI in Agriculture

Moreover, AI plays a role in crop monitoring and management by utilizing image recognition technologies automating tasks like weeding and harvesting. It also plays a role in climate prediction and risk management by providing weather forecasts and assessment models for potential risks. Smart irrigation systems powered by AI optimize water usage by analysing soil moisture levels along with other factors. The impact of AI extends beyond the fields. It influences supply chain optimization, livestock monitoring, market intelligence gathering well, as labour optimization—revolutionizing the entire agricultural ecosystem. As artificial intelligence continues to develop its role, in the sector is on the verge of bringing about advancements promoting sustainability, resilience and enhanced productivity.

3.4.1 Precision Farming

Sensor Technologies

Artificial intelligence integrates with sensor technologies such as drones, satellites and IoT devices to gather real-time data pertaining to soil conditions, crop health and weather patterns.

Data Analysis

Sophisticated AI algorithms process datasets to offer farmers insights that enable them to make precise decisions regarding irrigation, fertilization and pest control.

Figure 3.15 AI in Precision Agriculture

3.4.2 Crop Monitoring and Management

AI-powered image recognition tools diligently analyse satellite and drone imagery (refer to Fig. 3.16) to monitor crop growth accurately, identify diseases promptly and evaluate plant health.

Figure 3.16 AI-enabled Crop Management

Automated Weeding and Harvesting

Robotic systems driven by intelligence are equipped with computer vision capabilities that allow them to differentiate between crops and weeds seamlessly. This automation expedites the weeding process significantly. Additionally, harvesting robots utilize AI technology to identify crops for optimal harvesting.

Crop Disease Prediction and Management

AI models analyse historical and real-time data to predict the probability of disease outbreaks based on conditions and crop health status. Furthermore, drones equipped with AI powered cameras can swiftly detect signs of diseases or pests during their nascent stages - enabling farmers to take preventive measures.

Climate Prediction

Artificial intelligence (AI) examines weather data, satellite imagery and atmospheric conditions to provide weather forecasts, both short-term and long-term. This helps farmers plan their activities effectively based on the expected weather conditions.

Risk Management

By employing AI models, farmers can assess risks like weather events in advance. This allows them to take measures to protect their crops and livestock from any harm.

3.4.3 Smart Irrigation Systems

Sensor-based Monitoring

Modern practices leverage AI-enabled sensors installed in the fields to gather data on soil moisture levels, prevailing weather conditions and crop requirements.

Automated Irrigation

Using AI algorithms, gathered data is processed efficiently to optimize irrigation schedules accordingly (refer to Fig. 3.17). This ensures that crops receive an amount of water while minimizing water waste and improving irrigation efficiency.

Figure 3.17 AI-enabled Smart Irrigation System

3.4.4 Challenges in the Application of AI in Agriculture

While the integration of intelligence (AI) in agriculture offers benefits, it also poses several challenges that require careful consideration during implementation. Some key challenges include:

- **Limited Connectivity in Rural Areas:** Many agricultural regions lack robust internet connectivity, which hampers the seamless transmission of data necessary for the effective utilization of AI applications. The limited access to high-speed internet can affect real-time functionality of AI systems.
- **Limited Energy Infrastructure:** Many agricultural areas face challenges with consistent access to electricity, internet connectivity and other resources. The successful implementation of AI technologies may face obstacles due to the requirement for a steady power supply, or internet connectivity and other essentials.
- **Trust and Acceptance:** When it comes to trust and acceptance, farmers may have reservations about relying on AI recommendations. This hesitation can stem from concerns regarding transparency or a perceived disconnect between farming knowledge and insights generated by AI. Building trust is crucial for adoption as some farming communities might resist embracing AI due to preferences or a deep-rooted reliance on traditional farming methods.
- **Weather Dependency:** Weather dependency is another factor to consider. Agriculture heavily relies on accurate and timely weather data, which in turn affects the reliability of predictions and recommendations made by AI systems. Variability in weather patterns or sudden changes can impact the effectiveness of these predictions.
- **Dependency on Tech Companies:** As agriculture becomes increasingly reliant on AI, there is a risk of farmers becoming overly dependent on technology providers. This dependency could lead to issues if there are interruptions in service, changes in pricing models or discontinuation of support.

The integration of AI into agriculture holds potential for transforming farming practices, boosting productivity and contributing to sustainable agriculture. With advancements, we anticipate that the role of AI in agriculture will continue to evolve offering even more sophisticated solutions for addressing industry challenges.

3.5 APPLICATION OF AI IN EDUCATION

Artificial Intelligence is having an impact on transforming education by introducing learning experiences, improving administrative efficiency and enhancing student support (representative image in Fig. 3.18). Adaptive learning platforms use AI to tailor educational content based on individual student needs, creating a personalized and flexible learning environment. Intelligent tutoring systems provide virtual assistance by customizing teaching methods to suit the progress of students. Automated grading and feedback mechanisms save teachers time while providing insights into student's performance. Additionally, predictive analytics help identify learning challenges at a stage. Virtual classrooms and simulations powered by technologies like Virtual Reality (VR) and Augmented Reality (AR) create learning experiences. Language processing tools such as translation apps and language learning apps contribute to breaking down linguistic barriers.

Figure 3.18 Application of AI in Education

However, it is important to consider privacy concerns when integrating AI into education. Responsible collaboration between educators, technologists and policymakers is crucial for the implementation of AI while addressing the potential challenges. Striking a balance between the benefits of AI and ethical considerations is essential, for creating a landscape that is both technologically advanced and ethically sound.

3.5.1 Personalized Learning

Early Intervention Systems

AI-based analysis of student data helps identify learning difficulties or risks of dropout. This enables the implementation of intervention strategies aimed at providing support to struggling students.

Adaptive Learning Platforms

Adaptive learning platforms powered by AI analyse students' performance and customize content based on their needs. These platforms can suggest personalized learning materials, such as videos, articles and exercises, that align with each student's learning style. This approach enables students to learn at their own pace while receiving support in areas where they require assistance.

Intelligent Tutoring Systems

Tutoring systems driven by AI offer personalized guidance to students by addressing their questions, explaining concepts and providing feedback (representative image in Fig. 3.19). These systems adapt their teaching methods based on the progress of each student.

Figure 3.19 Smart Tutoring

3.5.2 Administrative Tasks

Enrolment Management

AI streamlines tasks like enrolment, scheduling and resource allocation to optimize the efficiency of educational institutions. By analysing large sets of educational data, AI can identify trends, assess institutional performance and contribute to decision-making processes.

Attendance Tracking

Facial recognition systems powered by AI automate attendance tracking, eliminating the need for manual recording. Additionally, AI can enhance campus security by monitoring surveillance footage for any unusual activity or potential threat.

Automated Grading and Feedback

AI algorithms can automate the grading process for assignments, quizzes and examinations—saving teacher's time while providing feedback to students. AI systems can provide personalized feedback by analysing students' work, aiding them in understanding their errors and enhancing their progress.

3.5.3 AI-based Language Processing Tools

AI-powered translation tools have the potential to bridge language barriers and facilitate communication and collaboration among students and educators, from different linguistic backgrounds. AI applications can support language learning by providing feedback on pronunciation, suggesting vocabulary and generating language exercises.

3.5.4 Challenges in Application of AI in Education

The implementation of AI in education comes with challenges that require consideration for responsible and effective use. Some of the key challenges are highlighted below.

- **Ethical Concerns** One significant challenge is ensuring ethical practices. AI systems in education often involve collecting and analysing student data making it crucial to protect privacy through security measures and compliance with data protection regulations. Additionally, there is a concern that AI algorithms may unintentionally reinforce existing biases in systems. It is important to address these issues to ensure fairness in grading and decision-making processes.

- **Teacher Training and Acceptance:** Another challenge lies in gaining acceptance from teachers and educational staff who may be resistant to adopting AI technologies. Convincing stakeholders about the benefits of AI while addressing concerns about job displacement or reduced autonomy can be obstacles to implementation and are key to successful implementation. Furthermore, providing training for teachers to effectively integrate AI tools into their teaching practices is essential.

- **Equity and Access:** We need to be mindful that AI-enhanced tools and technologies should be accessible to all students. The digital divide, which refers to disparities in access to technology and the internet, can widen the educational gap between students from socioeconomic backgrounds.

- **Pedagogical Effectiveness:** While we explore how AI can enhance learning outcomes, it is important that we conduct further research to understand how AI can best support pedagogical goals without replacing essential human elements in education.

- **Standardization of Regulatory and Policy Frameworks:** To ensure a smooth adoption of AI technologies in education, it is necessary to establish standardized regulations and policies that govern their implementation. Inconsistencies in implementation could hinder the use of AI in education.

3.6 APPLICATION OF AI IN TRANSPORTATION

In the transportation sector, AI is revolutionizing various aspects by improving efficiency, safety and sustainability. For instance, AI-driven traffic prediction and smart traffic lights that dynamically adjust signals based on real-time conditions help reduce congestion. Public transportation benefits from predictive maintenance and route planning enabled by AI. Additionally, intelligent transportation systems (refer to Fig. 3.20) and vehicle-to-infrastructure communication contribute significantly to safety and efficiency. Ride-sharing services utilize AI for dynamic pricing and personalized recommendations. They also prioritize safety by monitoring drivers. AI also plays a role in reducing environmental impact by optimizing fuel efficiency and keeping track of vehicle emissions.

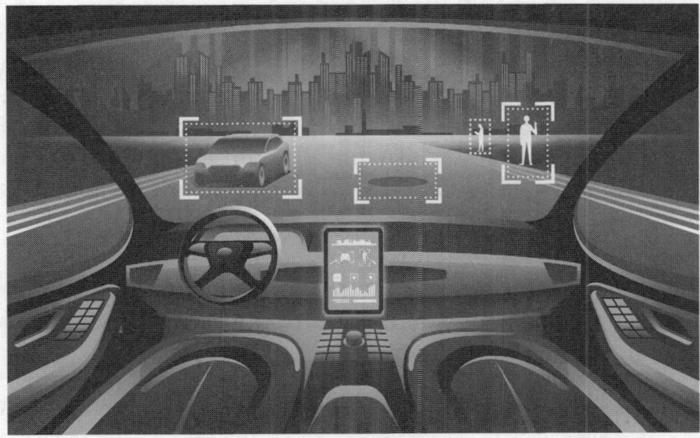

Figure 3.20 Application of AI in Transportation

However, integrating AI successfully in transportation comes with its set of challenges. These include regulatory considerations, ethical concerns and the need for upgrading infrastructure. To responsibly deploy AI technologies in the transportation sector it is crucial to ensure the security and privacy of data while establishing regulations. Now let us explore a few use cases that demonstrate how AI can be applied in transportation.

3.6.1 Traffic Management and Optimization

Traffic Prediction

AI algorithms analyse historical and real-time data to forecast traffic patterns accurately. This valuable information can then be used to optimize traffic signal timings, reroute vehicles and alleviate congestion (refer to Fig. 3.21).

Figure 3.21 Application of AI in Traffic Management and Optimization

Smart Infrastructure

AI can be seamlessly integrated into infrastructure components like roads, bridges and tunnels to enhance safety and efficiency. For instance, smart road signs, adaptive speed limits and automated toll collection are some features made possible through AI technology. Moreover, AI-powered traffic lights dynamically adjust signal timings based on real-time traffic conditions—reducing waiting times for drivers while improving traffic flow.

Route Optimization

Artificial intelligence algorithms analyse factors including passenger demand and traffic conditions to enhance transportation routes. This optimization process aims to improve efficiency and minimize travel times for commuters.

3.6.2 Ride-Sharing and Mobility Services

Dynamic Pricing

Dynamic pricing systems rely on real-time information to assess the demand for services compared to the availability of vehicles or seats. By monitoring these factors, the pricing algorithm can adapt fares accordingly striking a balance between supply and demand. During peak hours of demand, dynamic pricing may result in higher fares as a means of encouraging a more even distribution of demand among users. Conversely, off-peak hours might see lower prices to attract ridership and optimize service capacity. Adverse weather conditions or heavy traffic can lead to increased prices due to increased travel times or greater demand for alternative transportation options. Additionally, events, holidays or festivals can significantly impact transportation demands. Dynamic pricing effectively adjusts fares during occasions to accommodate increased user needs while ensuring that those willing to pay a premium can secure transportation services.

Surge Pricing

In ride-sharing services, surge pricing is a common strategy to adjust prices dynamically. When demand surpasses available drivers, prices can surge to encourage more drivers to join the platform and meet the increased demand.

Personalized Recommendations

One interesting application of AI systems is personalized recommendations for transportation options. These recommendations are based on user preferences, historical data and contextual information. For example, the AI system might suggest a route for a weekend drive or recommend public transportation options for a specific event. Over time the AI algorithm learns from the user's preferences and adjusts its recommendations accordingly.

3.6.3 Safety and Security

Driver Monitoring

When it comes to safety and security in transportation, driver monitoring systems play a role. These systems utilize AI and sensors like cameras and accelerometers to monitor aspects of a driver's behaviour. This includes detecting signs of fatigue, distraction, aggressive driving or impaired driving caused by substances such as alcohol or drugs. By capturing expressions and tracking eye movements through cameras, these systems can detect signs of drowsiness or distraction in drivers. Eye tracking technology proves valuable in assessing a driver's attentiveness to the road and identifying instances of drowsy driving. It can even prompt alerts or interventions when Advanced driver assistance systems (ADAS), as represented in Fig. 3.22, can monitor how a driver steers and stays in their lane. Driver monitoring systems can give drivers real-time alerts if they engage in unsafe behaviour.

Figure 3.22 Application of AI in Driver Monitoring

In the realm of transportation, driver monitoring often goes hand-in-hand with fleet management systems. This allows transportation companies to keep an eye on their driver's overall performance, pinpoint areas that need improvement and implement tailored training programs.

Security Surveillance

AI-powered video analytics bolster security at transportation hubs by identifying activities and potential security threats. The main objectives of security surveillance in transportation are to deter criminal activities, ensure passenger safety and respond swiftly to emergencies. Video cameras are strategically placed across locations including transportation hubs, stations, terminals and vehicles to provide monitoring. License plate recognition systems are employed to track and record the license plates of vehicles entering or leaving transportation hubs, parking lots or other critical areas.

3.6.4 Challenges in the Application of AI in Transportation

- **Reliability and Robustness:** Ensuring the reliability and robustness of AI algorithms is a challenge - especially when dealing with unpredictable transportation environments.

These algorithms must be adaptable to scenarios while effectively handling real-world complexities.

- **Human-Machine Interaction:** One of the challenges we face is how AI systems interact with humans who are driving, walking or riding as passengers. It is important to design interfaces and communication methods that are intuitive and promote collaboration. This is crucial for ensuring safety and gaining user acceptance.

- **Public Acceptance and Trust:** Gaining public acceptance and trust in AI-driven transportation can be difficult. We need to communicate to individuals that AI-powered traffic management systems are safe, reliable and beneficial. Transparent testing and validation processes are also essential in building trust.

- **Regulatory Frameworks:** Developing comprehensive and standardized regulations for AI in transportation poses a challenge. There is a need for clear guidelines addressing safety, privacy, liability, and ethical considerations to ensure responsible deployment.

3.7 CONCLUSION

In summary, the application of artificial intelligence (AI) has brought about a new era of innovation across various industries. It has transformed how we approach challenges and optimize processes. In healthcare AI-driven diagnostics and personalized treatment plans are revolutionizing patient care by providing accurate diagnoses and improving medical outcomes. The finance sector benefits from AI-powered algorithms used for risk assessment, fraud detection and algorithmic trading which enhance efficiency and security in transactions. In the retail sector, AI is revolutionizing how customers are served by offering recommendations efficiently managing inventory and optimizing supply chains. This is resulting in a customer-centric industry that responds effectively to individual needs.

In agriculture, AI plays a role in precision farming by providing insights into crop health helping to optimize resource allocation and ultimately improving overall agricultural productivity. The education sector also benefits from AI through personalized learning experiences, adaptive tutoring programs and streamlined administrative tasks. These advancements enhance the efficiency of processes and cater to the unique learning requirements of individuals. The transportation industry is undergoing a transformation as it integrates AI technologies. This integration has led to developments such as traffic management systems and enhanced safety measures.

While these advancements are impressive, challenges such as ethical considerations, privacy concerns, regulatory frameworks and societal acceptance need to be carefully addressed in each industry. Striking the right balance between technological innovation and responsible deployment is crucial to harness the full potential of AI. As these industries continue to embrace AI solutions, ongoing collaboration among industry stakeholders, policymakers, and the broader society will be essential to ensure that AI serves as a force for positive change, benefitting humanity while mitigating potential risks. The dynamic landscape of AI applications underscores the need for continual monitoring, adaptation, and ethical considerations to pave the way for a future where AI enhances the human experience across diverse sectors.

3.8 SUMMARY

- Across diverse sectors, AI applications are reshaping the way businesses operate, solve problems and interact with their stakeholders.

- The healthcare industry is at the forefront of incorporating AI technology, which has brought about changes in diagnostics, treatment plans and patient care.

- The widespread adoption of AI in the healthcare sector is driven by factors like:
  - The healthcare industry generates an abundance of data including images, patient histories and prescriptions.
  - Healthcare data has high variability complexity. It is also extremely intense a manual process to analyze and infer from medical data. So, automation is really needed to complement manual analysis.

- AI has also made an impact on the finance sector by revolutionizing financial operations. The applications of AI in finance are vast offering solutions that streamline processes, enhance decision making capabilities and improve customer experiences.

- AI has transformed the retail industry on various fronts by providing solutions that enhance operational efficiency and improve customer experiences.

- The application of AI in agriculture has ushered in a new era of innovation, fundamentally transforming traditional farming practices and addressing long-standing challenges.

- Artificial Intelligence is having an impact on transforming education by introducing learning experiences, improving administrative efficiency and enhancing student support.

- In the transportation sector AI is revolutionizing various aspects by improving efficiency, safety and sustainability. For instance, AI driven traffic prediction and smart traffic lights that dynamically adjust signals based on real-time conditions help reduce congestion.

- While these advancements are impressive, challenges such as ethical considerations, privacy concerns, regulatory frameworks and societal acceptance need to be carefully addressed in each industry. Striking the right balance between technological innovation and responsible deployment is crucial to harness the full potential of AI.

3.9 PRACTICE EXERCISES

3.9.1 Subjective Questions (10 Marks)

1. Explore the multifaceted applications of AI in healthcare, focusing on the role of AI in medical diagnosis. Discuss key areas such as radiology and imaging, retinal imaging,

pathology, dermatology, and cardiology. Provide examples of how AI algorithms contribute to accurate diagnoses and the early detection of various medical conditions.

2. Examine the integration of AI in the finance industry, with a specific focus on algorithmic trading. Discuss how AI analyses historical data, executes high-frequency trades, manages risks, and adapts trading strategies. Evaluate the advantages and challenges associated with AI-enabled algorithmic trading in the financial markets.

3. Delve into the role of AI in financial risk management, addressing areas such as credit scoring, fraud detection, and compliance with regulatory standards. Explore how AI models assess creditworthiness, detect fraudulent activities, and ensure compliance with anti-money laundering (AML) and know-your-customer (KYC) processes.

4. Examine the use of AI in customer service within the finance sector, including the deployment of chatbots, virtual assistants, and robo-advisors. Discuss how these AI-driven tools enhance customer support, automate routine tasks, and personalize financial planning services. Evaluate the challenges associated with handling sensitive information and maintaining regulatory compliance in AI-driven financial services.

5. Discuss how AI is transforming the retail industry, focusing on its impact on demand forecasting, inventory management, and personalization. Provide examples to support your discussion.

6. Examine the role of AI in inventory and store layout management in the retail sector. Discuss how dynamic inventory optimization, supply chain optimization, and smart store layouts contribute to operational efficiency. Support your answer with relevant insights.

7. Explore the applications of AI in providing a personalized shopping experience in the retail industry. Discuss the significance of personalized recommendations, visual search, and virtual try-on experiences. Use examples to illustrate how AI contributes to enhancing customer engagement.

8. Evaluate the impact of AI in agriculture, focusing on precision farming. Discuss how AI technologies, such as sensor integration and data analysis, contribute to optimal crop management. Provide examples to illustrate how precision farming practices have been enhanced by AI.

9. Analyse the application of AI in education, specifically in personalized learning experiences. Discuss how adaptive learning platforms and intelligent tutoring systems use AI to tailor educational content and provide customized support for students. Support your analysis with examples.

10. Examine the role of AI in transportation, focusing on traffic management and optimization. Discuss how AI contributes to traffic prediction, smart infrastructure, and route optimization. Provide examples to illustrate how AI enhances efficiency and reduces congestion in transportation systems.

11. Discuss the challenges associated with the application of AI in retail, agriculture, education, and transportation. Evaluate the common challenges and unique considerations for each sector. Use examples to support your discussion.

12. Explore the safety and security applications of AI in transportation, particularly in driver monitoring and security surveillance. Discuss how AI contributes to ensuring driver safety and enhancing security in transportation hubs. Provide examples to illustrate the effectiveness of AI in these areas.

3.9.2 Subjective Questions (5 Marks)

1. Explain the significance of AI in revolutionizing the healthcare sector, highlighting one specific application.

2. Discuss the role of AI in financial risk management, providing an example of how AI contributes to either credit scoring or fraud detection.

3. Examine the challenges associated with the integration of AI in healthcare, focusing on one concern related to data privacy and security.

4. Explore the transformative impact of AI in the e-commerce sector, emphasizing the role of AI-driven virtual assistants in enhancing customer interactions.

5. Describe how AI is reshaping the landscape of algorithmic trading in the finance industry, providing insights into the analysis of historical data.

6. Discuss one specific application of AI in personalized healthcare, such as its role in the early detection of diseases, and highlight its advantages.

7. Examine the challenges associated with AI in finance, focusing on the ethical concerns related to bias in algorithms and the lack of transparency.

8. Provide an overview of the diverse applications of AI in healthcare, emphasizing its role in pathology and histopathology.

9. Explain how AI is utilized in the retail industry to optimize inventory levels. Provide specific examples to illustrate the impact of AI on preventing overstocking or stockouts.

10. Discuss the role of AI-driven recommendation engines in the retail sector. How do these engines analyse customer behaviour and how does personalization contribute to increased sales? Support your answer with examples.

11. Examine the applications of AI in agriculture, focusing on precision farming. Provide examples to illustrate the impact of AI in this domain.

12. Evaluate the effectiveness of AI-powered intelligent tutoring systems in the education sector. Discuss how these systems provide personalized assistance to students and adapt to individual learning progress. Use examples to support your answer.

13. Discuss the challenges associated with the integration of AI in the transportation sector. Provide examples.

14. How do adaptive learning platforms tailor educational content to individual student needs, and what role do early intervention systems play? Support your answer with examples.

15. Examine the applications of AI in the retail industry's customer support. Provide examples to illustrate the effectiveness of these AI applications.

16. How do factors such as limited connectivity, trust and acceptance, and weather dependency impact the successful implementation of AI in agriculture?

17. Evaluate the role of AI in traffic management and optimization in the transportation sector. How do AI algorithms contribute to traffic prediction, smart infrastructure, and route optimization?

18. Discuss the ethical concerns associated with the application of AI in education. How does the collection and analysis of sensitive student data pose challenges, and what measures are essential to ensure responsible implementation? Provide examples to support your discussion.

3.9.3 Short Answer-type Questions (2 Marks)

1. Briefly explain one application of AI in the industry, highlighting its impact on customer experience.

2. Identify one challenge associated with the application of AI in transportation.

3. Name a specific area within financial risk management where AI contributes to decision-making.

4. Provide one example of how AI-powered virtual assistants enhance patient engagement in healthcare.

5. Highlight one concern related to data privacy and security in the context of AI applications in healthcare.

6. Explain the role of AI in personalized financial planning, mentioning one area where AI tailors recommendations based on individual preferences.

7. Name one industry challenge related to the integration of AI in finance, apart from regulatory compliance.

8. Define how AI contributes to supply chain optimization in the retail industry.

9. What is the role of AI in fraud detection in the retail sector?

10. Explain how AI enhances customer engagement in the retail industry.

11. How does AI contribute to climate prediction in agriculture?

12. Define the term 'adaptive learning platforms' in the context of AI in education.

13. What challenges does the agriculture sector face in terms of limited connectivity for AI applications?

14. Explain the concept of 'dynamic pricing' in the context of AI applications in transportation.

15. How does AI-powered video analytics contribute to security in transportation hubs?

16. Define the term 'early intervention systems' in the context of AI in education.

17. What challenges does the transportation sector face in terms of human-machine interaction with AI systems?

3.9.4 Multiple-choice Questions (1 Mark)

1. What is the primary role of AI in algorithmic trading in the finance industry?
   (a) Enhancing customer service
   (b) Analysing historical data and executing trades
   (c) Facilitating data privacy and security
   (d) Optimizing administrative tasks

2. What challenge is associated with AI applications in transportation?
   (a) Lack of standardized protocols
   (b) Overreliance on historical data
   (c) Fraud detection issues
   (d) Data privacy concerns

3. Which aspect of AI in finance involves assessing risk factors in real time and adjusting trading parameters?
   (a) Virtual assistants
   (b) High-frequency trading
   (c) Risk management
   (d) Algorithmic trading

4. In AI-powered virtual medical assistants, what is one area where they contribute to effective medication management?
   (a) Administrative tasks
   (b) Pathology
   (c) Medication reminders
   (d) Dermatology

5. What is a challenge related to the application of AI in healthcare?
   (a) Lack of standardized protocols
   (b) Overreliance on historical data
   (c) Data privacy and security concerns
   (d) Regulatory compliance issues

6. In personalized financial planning with AI, what does the technology tailor recommendations based on?
   (a) Regulatory compliance
   (b) Individual preferences and financial goals
   (c) High-frequency trading
   (d) Administrative tasks

7. What is the primary focus of AI-driven recommendation engines in the retail industry?
   (a) Enhancing security
   (b) Dynamic pricing
   (c) Personalization
   (d) Fraud detection

8. How does AI contribute to precision agriculture?
   (a) Predicting market trends
   (b) Monitoring traffic conditions
   (c) Enhancing crop management
   (d) Analysing customer behaviour

9. What is the primary function of AI-powered chatbots in retail?
   (a) Predicting weather conditions
   (b) Providing customer support
   (c) Dynamic pricing adjustments
   (d) Enhancing security

10. In education, what do adaptive learning platforms use AI for?
    (a) Predicting student behaviour
    (b) Customizing teaching methods
    (c) Analysing weather patterns
    (d) Managing traffic conditions

11. What is a common challenge in the application of AI in agriculture?
    (a) Predicting market trends
    (b) Limited connectivity in rural areas
    (c) Dynamic pricing adjustments
    (d) Smart store layouts

12. What does dynamic pricing refer to in the context of AI in transportation?
    (a) Adjusting inventory levels
    (b) Real-time pricing adjustments
    (c) Predicting weather conditions
    (d) Enhancing customer engagement

13. How do intelligent tutoring systems contribute to education?
    (a) Providing customer support
    (b) Automating grading processes
    (c) Customizing teaching methods
    (d) Analysing traffic conditions

14. What is the role of AI in driver monitoring in transportation?
    (a) Analysing customer behaviour
    (b) Enhancing crop management
    (c) Monitoring driver behaviour
    (d) Providing real-time updates

15. What is the primary challenge in AI applications in transportation related to human-machine interaction?
    (a) Limited connectivity
    (b) Weather dependency
    (c) Regulatory frameworks
    (d) Gaining public trust

16. In retail, what does the term "visual search" refer to in the context of AI?
    (a) Analysing customer behaviour
    (b) Searching for products using images
    (c) Predicting market trends
    (d) Adjusting inventory levels

3.10 ANSWER KEYS

3.10.1 Multiple-choice Questions

1. (b)	**2.** (a)	**3.** (c)	**4.** (c)	**5.** (c)	**6.** (b)	**7.** (c)	**8.** (c)
9. (b)	**10.** (b)	**11.** (b)	**12.** (b)	**13.** (c)	**14.** (c)	**15.** (d)	**16.** (b)

Bias and Fairness in AI Systems

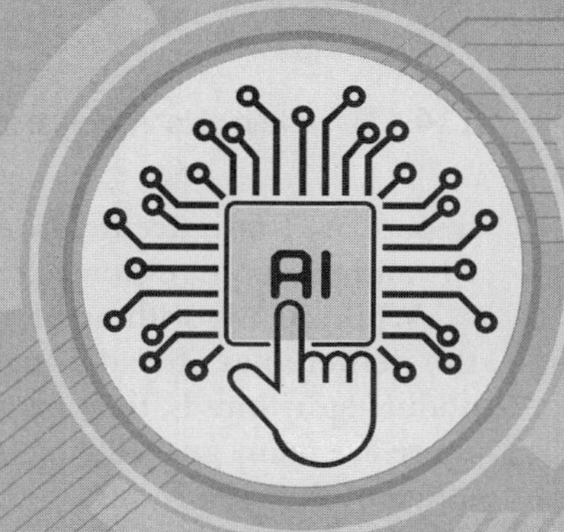

4.1 INTRODUCTION

As discussed earlier many AI Programs have beaten the best of humans in the business. Twenty-five years back 'Deep Blue' defeating Garry Kasparov or AlphaGo beating Lee Sedol in the game of GO have certainly caught our attention. This is sure to evoke mixed feelings, one that is of awe and bewilderment, another may be of fear (AI taking over the human race, singularity) and one another may be that of scepticism. Well, the breakthroughs of AI programs are not only limited to games but also to education, healthcare, transportation, agriculture and many other sectors. As this AI system is getting integrated with many daily activities, there are a few non-tangible issues that are cropping up like Ethics in AI, Fairness and Bias in AI and so on.

Before, we jump into AI of Ethics, let's revisit some general concepts of ethics. Ethics is a branch of philosophy that deals with moral principles, values and general rules of individual or group behaviour. Basically, this involves the study and exploration of what is considered right and wrong, good and bad, and the moral implications of human actions. Ethics provides a framework for making decisions and determining the appropriateness of behaviour within various contexts. Ethics is of course very different from legal requirements. There are formal rules embodied in terms of law. If there is a deviation from observing the law, it is enforceable and one may face consequences like fine, imprisonment, etc. In contrast, ethics is more like a moral compass, one is not going to face any legal consequence. His or her image or reputation in society may be affected at most. There are several such things which are not unethical but not legal. For example:

- Many companies may be knowingly selling products that can be harmful to consumers or the environment, ranging from addictive social media algorithms and products designed for planned obsolescence (Deliberately using techniques to shorten life span) to traditional harmful products like cigarettes.
- Taking credit for someone else's work or ideas. Plagiarism, even in casual settings, can be a form of intellectual theft and undermines the originality and creativity of others.
- Creating and spreading misinformation, even if not done with malicious intent.

Now let's first understand, what is ethics in designing a software product or generic software development? Ethics in IT refers to the principles and standards of conduct that guide professionals in the Information Technology (IT) industry. It involves making decisions and taking actions that are morally and socially responsible. Key aspects of IT ethics include respecting privacy, protecting data security, ensuring fair use of technology and considering the societal impact of IT advancements. Adhering to ethical principles helps build trust, maintain integrity and promote responsible behaviour within the IT community.

While AI Algorithms have matured significantly in the last decade, algorithms (one of the deliverables of traditional IT) have been present for much longer times. A quick recap about algorithms is, they are a sequence of instructions to complete a task. For any given task, there are many given algorithms at our disposal. We choose the optimal ones, based on how efficient they are in terms of computational time, memory usage, etc. We cannot overlook the fact that machine learning or AI algorithms have been designed differently from conventional IT Programs.

The below table gives a brief summary of traditional software *vs* AI-based software.

Feature	Traditional Software	AI Software
Development	Rule-based, pre-programmed	Data-driven learning from data
Learning	Limited, requires manual updates	Continuous, self-improves with new data
Adaptability	Low, inflexible to changes	High, adjusts to new situations
Generalizability	Limited to specific tasks	High, applies to broader problems
Explainability	Easy to understand, logic-driven	Often a 'black box,' difficult to explain decisions

Let's now explore, some of the issues concerning AI-based software, model or algorithm in more details. Some of the issues are as follows:

1. AI algorithms deal with much more amount of data than traditional algorithms. Because of the amount of data required, often data requires to be collected from scratch. This data may introduce **bias** and hence may question the **fairness** practice adopted in the design and development of the algorithm/model/system (detailed examples are discussed later). This also pronounces the existing issue of **data security and privacy.** Let's look at a couple of issues related to data privacy.

   a. **Informed Consent and Data Collection:** One major concern involves the collection of personal data without clear, informed consent from individuals. AI systems often rely on vast datasets to train and operate effectively. Ensuring that individuals understand how their data will be used and obtaining explicit consent is essential to respect privacy.

   b. **Surveillance and Individual Privacy:** AI technologies, particularly in surveillance systems, can raise concerns about the infringement of individual privacy. Striking a balance between the benefits of enhanced security and the protection of individual privacy rights is a complex challenge that requires careful regulation and oversight. Any AI model needs a lot of training data and we the users are providing all this data based on our searches, actions, clicks, likes, comments, etc. Many applications ask us to share location and allow access to microphone, camera, etc. While these come with standard disclaimers, hardly users go through them line by line. As a result, many personal data is available with these applications. On one side, they may be abused by the organizations directly, on the other side, if this data is not safely kept, that may be abused by some third party.

2. The working of the algorithm is not completely known to the designer. Because AI Algorithms are not explicit. For some of the algorithms, the internal working

is somewhat understandable and is classified as white box, glass box, etc. The developer or the designer knows how the model parameters are to be determined to best fit the data. However, that knowledge does not make the algorithms completely **transparent**. Hence machine learning algorithms at best can be thought of as meta-algorithms. An Algorithm that can write an algorithm. Here comes a dilemma. In a traditional IT scenario, if someone writes malicious code, the developer will be held accountable. Based on the nature of the maliciousness, it can be legally punishable too. However, in case of AI algorithms in spite of applying sound design principles, the algorithms may produce unwanted results. In this case, who would take the **accountability**? Hence, AI algorithm increases the risk of not being transparent. As a result, **explainability**, **Interpretability**, and **causality** for AI Models have become a very active research area. On the other hand, some models can produce superlative results but are black box models. In this case, how one will ascertain that the outcomes are **trustworthy**?

3. Often, the models are published and made available to the community. Later, we will discuss that even the models expose some risks.

Let's explore some examples. Self-driving cars are an excellent manifestation of AI. It has been tested over many scenarios, road conditions, etc., however, let's say it detects a person lying wounded on the side of the road, will it stop to call the ambulance or will it continue driving? Another example may be, if it sees an animal in the road suddenly appearing, will it break sharply, which may overturn the car and damage itself or will it hit the animal?

Let's take an example of AI System being used in the US Judiciary. Correctional Offender Management Profiling for Alternative Sanctions or COMPAS is a tool, which is used to find the probability/likelihood (can be used equivalently in a loose sense) of an offender becoming a repeat offender (recidivism). This is also applicable to substance usage. Suppose the algorithm had more false positives (i.e., it predicted for a repeat offence, but in reality there was no repeat offence) was high for coloured individuals. There are many such infamous examples, Figure 4.1 is an example where a person with criminal history but white skin was treated as low risk compared to a first-time offender with black skin.

Figure 4.1 Affect of demographics on Risk Profile

DO YOU KNOW

AI Algorithms are sometimes accused of suffering of Hungry Judge Effect. The hungry judge effect is a hypothesis that suggests that judges are more likely to make harsher rulings when they are hungry. This effect has been observed in a number of studies, but it is still a subject of debate among researchers.

4.2 ETHICS IN AI

Ethics in AI refers to the principles, guidelines and standards that govern the responsible and ethical development, deployment, and use of artificial intelligence systems. It involves addressing the moral and societal implications of AI technologies to ensure that they are developed and used in ways that align with human values and adhere to ethical norms.

Below we discuss some of the key areas as far as AI in Ethics is concerned:

- **Fairness:** Ensuring that AI systems are unbiased and treat all individuals or groups fairly, without discriminating based on factors such as race, gender, or socioeconomic status.

- **Transparency:** Promoting transparency in the development and decision-making processes of AI systems, making it clear how algorithms operate and arrive at their conclusions. As highlighted previously, interpretability, explainability and causality are some of the factors in achieving transparency.

- **Accountability:** Establishing mechanisms to assign responsibility when AI systems cause harm or make incorrect decisions and ensuring accountability for developers, organizations and users.

- **Privacy:** Safeguarding individuals' privacy by respecting and protecting their personal data when collecting, processing, or using information within AI systems. This is an issue being looked at very seriously. There is a trade-off here. On one hand, if corporates, governments, NGOs, academic institutes and healthcare providers do not publish open datasets, research will become very difficult, on the other hand, this can raise serious data privacy issues.

- **Security:** Implementing measures to secure AI systems against potential vulnerabilities, hacking, or misuse, preventing unauthorized access or manipulation.

- **Inclusivity:** Considering the diverse perspectives and needs of different stakeholders, ensuring that AI technologies benefit society as a whole rather than exacerbating existing inequalities. There is some overlap in the concept of Inclusivity with that of transparency.

- **Sustainability:** Evaluating the environmental impact of AI systems and working towards minimizing their energy consumption and carbon footprint. AI Models, especially deep learning models have billions of parameters and need as much to be trained on. Hence the computational energy demand is way too high.

- **Robustness and Reliability:** The robustness of an AI model refers to its ability to maintain accurate and reliable performance across diverse and challenging conditions. Achieving robustness is crucial for enhancing the model's resilience and ensuring its reliability in a wide range of practical settings.

Ethics in AI is an evolving field as technology advances and discussions surrounding ethical considerations continue. Many organizations, researchers, and policymakers are actively engaged in developing ethical frameworks and guidelines to guide the responsible development and use of AI. There is a related concept, which is being discussed called Responsible AI. Issues like robustness and reliability may be more relatable with responsible AI, than Ethical AI.

4.3 BIAS AND FAIRNESS IN AI SYSTEMS

AI seems to be all-powerful in recent days, but what if the models are biased? Bias in AI systems is a critical concern that has garnered significant attention in recent years. It refers to the presence of systematic and unfair discrimination in the outcomes or predictions generated by artificial intelligence algorithms. Bias can manifest in various forms, including but not limited to race, gender, age, socioeconomic status and more. Understanding bias in AI is essential because it has profound implications for individuals, society and the adoption of AI technologies.

DO YOU KNOW?

In 2016, the inaugural international beauty competition, relied on 'machines' for judging the competition. The machine/program was given the responsibility to assess contestants based on objective criteria such as facial symmetry and wrinkles to determine the most attractive participants. Following the launch of *Beauty.AI* earlier that year, over 6,000 individuals from over 100 countries eagerly submitted their photos, anticipating that artificial intelligence, driven by sophisticated algorithms, would identify those whose faces most closely embodied the concept of 'human beauty.' However, upon receiving the results, the organizers were disheartened to discover a troubling commonality among the winners: the robots displayed a clear bias against individuals with darker skin tones.

Let's look at some more historic incidents, which questioned the fairness of the AI Models. The first incident requires a little bit of prelude. Word Embedding is a highly cited work by Google and other teams. It establishes the similarity between two words and this similarity is not confined to trivial examples like linking 'no' with 'not' but can identify that the similarity between man and women is the same as that of king and queen. The word embedding model was doing miracles in terms of relating words. However, in one such example, it is observed that the relationship between ladies and earrings is the same as that of nephews and geniuses. This was taken as sexist, as women were getting related to ornaments and boys with intelligence. In 2016, researchers from Boston University and Microsoft wrote an article titled 'Man is to Computer Programmer as Woman is to Homemaker', where they discussed about the somewhat disturbing effects of gender stereotypes.

Another such incident drew quite a lot of attention in 2018. Amazon was using an AI model which was shortlisting candidates for software engineering positions. It was found that the model was penalizing resumes which has mentioned about anything to do with women. You can understand the hue and cry it would have caused. Many of these biases happen because of the training data that they are fed on. In the following section, we discuss about different types of bias.

4.3.1 Types of Bias

Data Bias

Data bias occurs when the training data used to build AI models is unrepresentative or skewed. If the data predominantly contains information from a specific group or exhibits historical prejudices, the AI model can inherit these biases. For example, if a facial recognition dataset contains primarily images of lighter-skinned individuals, the model may perform poorly on darker-skinned individuals. Some more examples of Data Bias are as follows:

- **Gender Bias in Hiring:** If historical hiring data predominantly reflects one gender over another due to past discriminatory practices, algorithms trained on this data may inadvertently favour one gender in current hiring decisions, perpetuating the bias.
- **Racial Bias in Policing:** If police arrest records are used to train predictive policing models, they may result in over-policing in certain neighbourhoods, as arrests in those areas are more frequent, leading to racial bias in law enforcement. A case regarding recidivism has been discussed already in this context.
- **Age Bias in Loan Approvals:** If a bank's historical data shows a preference for approving loans to younger applicants, AI models trained on this data may discriminate against older applicants seeking loans.
- **Geographical Bias in Healthcare:** Healthcare data may be biased towards urban areas due to better healthcare infrastructure and more comprehensive data collection. This can lead to disparities in healthcare resource allocation in rural or underserved regions.
- **Language Bias in Sentiment Analysis:** Sentiment analysis models trained in a specific language may struggle to understand and interpret sentiments accurately in languages they were not trained in, leading to language bias.
- **Healthcare Diagnosis Bias:** Diagnostic AI systems may provide different accuracy rates across demographic groups due to variations in data quality or representation in the training data, leading to healthcare disparities.

Algorithmic Bias

Algorithmic bias is introduced during the design and development of AI algorithms. It can result from flawed assumptions, model architectures, or feature selections. Biased algorithms may amplify existing prejudices or fail to correct them. For instance, a language translation model might produce gender-biased translations due to gender-neutral languages being misrepresented.

Societal Bias

Societal bias is deeply ingrained in societal values and can influence the decision-making of AI systems. It is often the most challenging bias to identify and mitigate. It goes beyond individual biases and is ingrained in the social, economic and political structures, institutions and norms.

- Societal bias can influence the way data is collected. For example, if there is a bias against a particular racial or ethnic group in society, it may affect the data collected about that group. This can result in the under representation or misrepresentation of certain groups in datasets.

  Schwemmer *et al.,* focussed particularly on detecting gender bias in object recognition systems. They have employed a commercial image recognition system and prepared datasets consisting of US congressmen and public images from the social media platform X (Earlier known as Twitter). Some of the observations from the research article are as follows.

- Three times more annotations related to physical appearance are found in images of women.

- Furthermore, substantially lower recognition rates are observed for women in comparison with men in images.

Figure 4.2 is a revealing finding, where in an exercise, vector embedding of words have been prepared using Wikipedia Corpus. Then similarity of some job roles is calculated with respect to the vectors corresponding to 'male' and 'female'. Occupations like "maestro", "skipper", "philosopher" architect", "financier", "pilot" are more similar with the 'he' vector (Circles). Occupations like "homemaker", "nurse", "receptionist", "librarian", "nanny", "stylist", "housekeeper", "counsellor" are more similar to the she vector (Triangle).

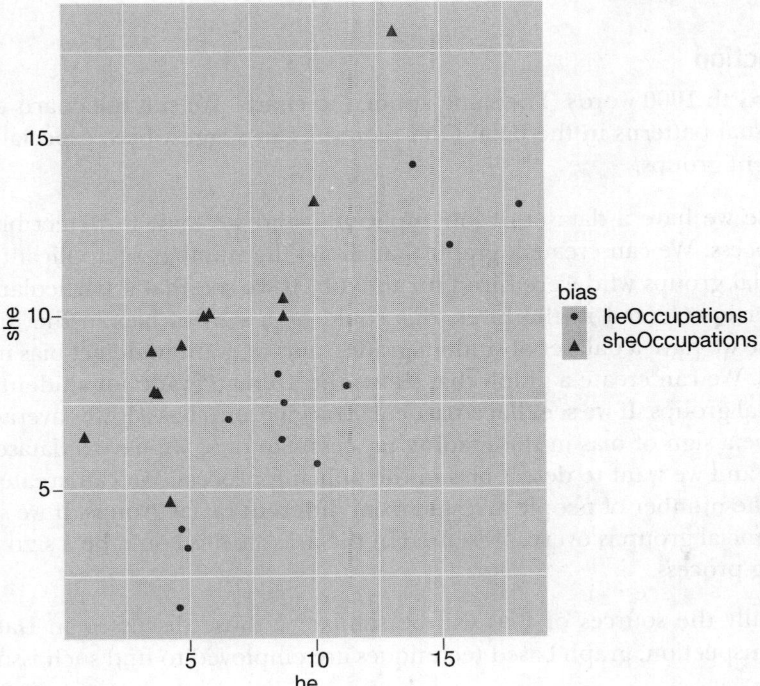

Figure 4.2 Top labels for images of men vs. women from social media

Some immediate finding of gender stereotypes follows:

- More professional work role type words are associated with men, for example, businessperson, spokesperson, and white collar worker while for women it is limited to television presenter.

- Gentleman appears much more formal as compared to 'Girl' as seen in the image labels.

- More physical appearance related to words are present in women.

4.3.2 Measuring and Detecting Bias

One of the fundamental ways of detecting bias is by analysing the data.

Data Review

The first step is to check the data for any variations or differences between groups. If one group is underrepresented or treated unfairly, it might indicate bias. Below are some of the examples,

- In a hiring AI system, if there's a lack of resumes from certain gender groups in the dataset, it indicates data bias. In a loan approval model, if most of the historical data is from one neighbourhood, it might not be representative of all potential borrowers. If a medical AI system has data only from one region, it may not generalize well to patients from other regions.

Visual Inspection

A picture is worth 1000 words. The same applied to charts. We can use charts and graphs to look for unusual patterns in the data. Odd patterns can suggest bias, especially when comparing different groups.

- Suppose we have a dataset of job applicants, and we want to detect bias in the hiring process. We can create a graph that shows the number of applicants from different racial groups who were hired for the job. If we see that a particular racial group is underrepresented in the hires, this could be a sign of bias in the hiring process. Suppose we have a dataset of student grades, and we want to detect bias in the grading process. We can create a graph that shows the average grade for students from different racial groups. If we see that a particular racial group has a lower average grade, this could be a sign of bias in the grading process. Suppose we have a dataset of criminal arrests, and we want to detect bias in the policing process. We can create a graph that shows the number of people arrested from different racial groups. If we see that a particular racial group is overrepresented in the arrests, this could be a sign of bias in the policing process.

- Essentially the sources of Bias will be similar to those discussed in Data Review. In Visual Inspection, graph-based techniques are employed to find such issues.

Check Fairness

We can calculate fairness metrics related to our specific problem. For example, we can try and see if AI system is making fair decisions for different groups. Fairness metrics are mathematical measures that can be used to assess the fairness of machine learning models. Fairness or parity can be found on many dimensions, one of them being demographic parity.

Let's find out how we can compute the demographic parity metric:

- **Step 1:** Calculate the prevalence of the positive outcome in each demographic group. For example, if we are looking at a loan approval model, we would calculate the percentage of loans that are approved for each demographic group, such as gender, race, and ethnicity.

- **Step 2:** Compare the prevalence of the positive outcome across demographic groups. If the prevalence of the positive outcome is equal in all demographic groups, then we say that the model has achieved demographic parity. Otherwise, we say that the model has not achieved demographic parity.

Here is a concrete example:

- Suppose we have a loan approval model that is trained on a dataset of historical loan applications. We want to assess the fairness of the model, so we calculate the demographic parity metric.

- We may find that the prevalence of loan approval is 70% for white males, 65% for white females, 60% for black males, and 55% for black females.

- This means that the model is not achieving demographic parity, because the prevalence of loan approval is not equal across all demographic groups. Black males and black females are less likely to have their loans approved than white males and white females.

Compare Outcomes

It can be investigated if protected groups are affected more than others. If there's a significant difference, it could be a sign of bias.

4.4 TRANSPARENCY IN AI SYSTEMS

As we are increasingly becoming dependent on decisions made by AI, we need to understand the decision-making process of these models to some extent, to have trust in the system. Well, if an oncologist wants to use a computer vision program, which can detect cancer by looking at the MRI Scan, he definitely needs to understand how the model is arriving at this decision. A bank can use an AI-based program for the loan, however, if an application is rejected based on the same, a higher authority of the bank, or the central bank, or some other regulatory body may want to know the reason for the same.

Transparency in an AI system refers to the clarity and openness in understanding how the system operates, makes decisions and processes information. It involves making the inner workings of the AI system accessible and understandable to users, developers and other stakeholders. Transparency is crucial for building trust, ensuring accountability and addressing concerns related to bias, fairness and ethical considerations in AI.

One thing, that needs to be kept in mind is, as we explore all these concepts related to Ethical AI, like fairness, transparency, accountability, privacy, etc., often these concepts overlap. For example, you can know about the fairness of a model, only when the model is transparent. Let's look at the key aspects related to transparency.

- **Explainability:** The ability to explain how the AI system arrives at a particular decision or recommendation. This involves providing insights into the factors and features considered by the algorithm and the reasons behind its output. Explainable AI (XAI) techniques aim to make complex models more interpretable. For instance, in healthcare, an XAI system might explain the factors considered in predicting a patient's diagnosis, helping physicians and patients comprehend the rationale behind the AI's recommendations and fostering trust in the decision-making process.

DO YOU KNOW?

On March 18, 2018, an Uber self-driving car struck and killed Elaine Herzberg, a pedestrian, in Tempe, Arizona. It was the first known death of a pedestrian caused by a self-driving car. The Uber car was operating in autonomous mode with a human safety backup driver sitting in the driving seat. The safety driver, Rafaela Vasquez, was streaming a television show at the time of the crash. Herzberg was pushing a bicycle across a four-lane road when the Uber car hit her. The car was traveling at about 40 mph at the time of the impact. Herzberg was taken to the hospital, where she died of her injuries. The National Transportation Safety Board (NTSB) investigated the crash and found that the Uber car's self-driving system was at fault. The NTSB also found that the safety driver's distraction was a contributing factor. In addition to the NTSB investigation, the Arizona Department of Public Safety also conducted a criminal investigation into the crash. In 2019, the Maricopa County Attorney's Office announced that it would not charge Uber with any crimes. However, the office did charge Vasquez with negligent homicide. Vasquez was acquitted of the charge in 2021.

- **Visibility into Data and Training:** Transparency involves disclosing the data used to train the AI system, including information about the sources, quality and representativeness of the data. Knowing the training data helps in understanding potential biases and limitations in the system.
- **Algorithmic Transparency:** Understanding the algorithms and models employed by the AI system is crucial. This includes transparency about the architecture, parameters and optimization methods used in machine learning models. Open-source AI frameworks contribute to algorithmic transparency.
- **Decision-Making Processes:** Transparent AI systems provide clarity on how decisions are made, especially in critical applications like finance, healthcare, or criminal justice. Users should know the criteria considered and the process leading to a specific outcome. Again, decision-making process has some overlap with XAI. The decision-making process looks at the entire sequence of the task, whereas XAI focuses on the model.
- **Ethical Considerations:** Transparency involves addressing ethical considerations in AI development and deployment. This includes being transparent about the ethical principles followed, potential risks and the measures taken to mitigate biases and unfairness.
- **User Interface Design:** A transparent AI system should have an intuitive and user-friendly interface that allows users to interact with and understand the system's functionalities. Clear communication of system capabilities and limitations is essential.
- **Documentation and Reporting:** Providing comprehensive documentation about the AI system's design, functionality and performance metrics contributes to transparency. Regular reporting on the system's performance, updates and any incidents is also important.
- **Compliance with Regulations:** Transparent AI systems adhere to legal and regulatory requirements related to data privacy, security and fairness. Compliance with standards and regulations contributes to accountability and transparency.

While transparency in AI systems is generally considered beneficial, there are some potential negatives to consider.

4.4.1 Security Risks

- **Data Exposure:** Providing too much detail in transparency reports may inadvertently expose sensitive information or vulnerabilities in the system, leading to potential security risks.
- **Model Exploitation:** Detailed insights into the model's architecture and decision-making processes could be exploited by malicious actors to manipulate or attack the system.

4.4.2 Misinterpretation by Users

- **Complexity Overload:** In-depth technical details may overwhelm non-expert users, leading to misinterpretation or misunderstanding of the system's functioning.

- **False Sense of Certainty:** Transparency does not guarantee users' ability to predict every decision accurately and overestimating this capability may result in a false sense of certainty.

4.4.3 Competitive Disadvantage

- **Intellectual Property Concerns:** Full transparency might expose proprietary algorithms or unique features, providing competitors with insights that could impact a company's competitive advantage.
- **Strategic Decision Revealing:** Complete transparency may disclose strategic decisions or business intentions, potentially affecting a company's market position.

It's essential to strike a balance in transparency, providing meaningful insights without compromising security, user comprehension, or a company's competitive edge. Each application and context may require a tailored approach to transparency based on these considerations.

4.5 ACCOUNTABILITY OF THE AI SYSTEMS

Many small and big decisions are being taken by AI Systems, but what happens if something goes wrong? Before proceeding further, let us get introduced to OECD.AI. OECD stands for the Organisation for Economic Co-operation and Development. It is an intergovernmental economic organization with 38 member countries, founded in 1961 to stimulate economic progress and world trade. OECD.AI works under the broad umbrella of the OECD in the area of AI. Some of OECD's observation can be seen below.

The concepts of accountability, responsibility and liability are interconnected yet distinct, and their interpretations may vary across cultures and languages. Broadly, 'accountability' entails an ethical, moral, or similar anticipation, such as those outlined in management practices or codes of conduct, guiding individuals or organizations in their actions. It enables them to elucidate the reasons behind decisions and actions taken. In cases of unfavourable outcomes, accountability involves taking corrective measures to ensure improved results in the future. On the other hand, 'liability' primarily addresses adverse legal consequences resulting from the actions or inaction of an individual or organization.

'Responsibility' also encompasses ethical or moral expectations and finds application in both legal and non-legal contexts to denote a causal connection between an actor and an outcome. Considering these nuances, the term that most aptly captures the underlying principle is 'accountability.' In this context, accountability signifies the expectation that organizations or individuals will guarantee the proper functioning of AI systems they design, develop, operate, or deploy throughout their lifecycle. This alignment should adhere to their respective roles and relevant regulatory frameworks. Demonstrating accountability involves actions and decision-making processes, such as furnishing documentation on pivotal decisions made during the AI system lifecycle or permitting auditing when warranted.

4.6 PRIVACY AND DATA PROTECTION CONCERNS

Privacy and data protection concerns in the context of artificial intelligence (AI) are critical aspects that demand careful consideration. Let's look at some of the examples where the collectors thought they had taken precautionary measures which proved to be otherwise.

The first incident reported was in Massachusetts in 1990. A Government agency was keeping record of all hospital visits. They decided to keep the dataset available for research after duly removing identifying fields like name, address, social security number, etc. They however kept the date of birth, zip code and gender for the sake of summary statistics. Later, a researcher from MIT showed that using this information 87% of the individuals could be identified.

Several AI literatures talk about the Netflix Competition in 2006 as it is a landmark event as far as competitive data science is concerned. This dataset was prepared and shared with the general public after a lot of due diligence in removing all possible identifiable information. Even for this dataset, later some researchers from the University of Texas at Austin could identify some people within just two weeks of publishing the dataset. It may be noted that since 1988, there has been a law in vogue called as Video Privacy Protection Act, which says if any rental company makes the video rental record of any customer public, it is liable for a 2,500 $ fine.

Let's explore the issue of privacy, with the following example of a fictitious hospital in Kolkata:

Name	Age	Gender	Locality	Ailment
Ram	53	M	Maniktala	Heart Disease
Dia	61	F	Salt Lake	Anaemia
Rajen	55	M	Shyam Bazar	Stroke
Anita	48	F	Rajarhat	Typhoid
Nur	62	M	Garia	Dengue
Minati	67	F	Park Circus	Mental Disorder

If now this dataset is released by removing the names, if someone just knows Minati's age and gender, they can find out that she has a mental disorder, which may not be a good thing to happen.

An initial solution of the same is k-anonymity. The idea is even if we can filter by some of the dimensions or attributes, we will not be able to single out individual cases, i.e., there will be at least k individuals satisfying the criteria. For example, instead of the exact age, if we store an age range, there will be two records that are female and have ages between 61–70. Hence, this will be 2-anonymous.

More details of these can be found in the book named 'The Ethical Algorithm'. Another interesting thing is that while there are definitely concerns about sharing the data about let's say, patients or customers, sharing models should not be of any concern at all. Because the models should just expose the architecture and values of different parameters only. However, it turns out that some knowledge of the data also gets leaked through these parameters of the models.

In the same book, the authors have highlighted a very interesting dilemma. Data collection and sharing are necessary for the advancement of knowledge and research, but would anyone be interested in sharing their data, if there is a privacy risk or there is an apprehension that the results may be used against them? Let's say Sunil participates in a survey, which later on establishes a correlation between drinking and increased blood pressure (maybe this survey is a little old when this correlation was not in public knowledge). Now, as this comes to common knowledge and an insurer knows that Sunil drinks, he/she is surely going to charge a hefty insurance premium from Sunil. This argument may put a question mark on all data sharing. Because the data is anonymized and correlation is established at an aggregate level and still a person like Sunil stands to lose from such a study. An intelligent reader may ask this question, participating in a survey is voluntary. Sunil may not have participated in the survey in the first place. But if we think for a while, we would be conveniently figuring out that, even if Sunil had not participated in the survey, just by having the knowledge that Sunil drinks and drinking causes blood pressure, the insurer any way would have charged a higher premium for his insurance. Frankly, if this is the expectation from AI algorithms that the results will not bring any kind of economic losses, this cannot be ensured. However, what can be ensured is Differential Privacy. Loosely it means, one's differential privacy is preserved, if the inclusion or exclusion of his record in the study, does not change the outcome of the same. More formally it can be stated as follows:

The primary goal of differential privacy is to allow the gathering of useful insights from data while protecting the privacy of individuals in the dataset. It provides a mathematical definition of privacy that quantifies the impact that the inclusion or exclusion of a single individual's data would have on the outcomes of the analysis.

The concept of differential privacy was introduced by a set of theoretical computer scientists in 2000, which actually won them the coveted Godel Prize. One other thing about differential privacy is randomness. When it is used in differential privacy, some amount of noise or disturbance is added to the individual records. This ensures, that the conclusions drawn are not inaccurate as well as the individual records cannot be found out through reverse engineering.

A differential privacy parameter of 1.5, often denoted as 'epsilon' (ε), represents the level of privacy protection in a differentially private system. The epsilon value is a key parameter in differential privacy, and lower values indicate stronger privacy protection.

In the context of differential privacy:

A smaller epsilon value corresponds to a higher level of privacy protection. A larger epsilon value allows for more noise or randomness to be introduced, potentially offering less privacy protection. The epsilon value is used to control the trade-off between privacy and utility in data analysis. A common interpretation is that smaller epsilon values provide a stricter privacy guarantee but may result in less accurate or useful outcomes.

For example:

- An epsilon value of 0.1 indicates a very strong privacy guarantee, but it may lead to noisier and less accurate results.
- An epsilon value of 1.0 provides a reasonable balance between privacy and utility.
- An epsilon value of 10 allows for more flexibility in the analysis but may provide weaker privacy protection.

It's important to note that the interpretation of epsilon can vary based on the specific differential privacy mechanism and the context in which it is applied. Differentially private algorithms and systems often aim to keep the impact on privacy, measured by epsilon, within a predetermined acceptable range. The choice of epsilon is a critical aspect of designing and implementing differentially private systems to achieve the desired level of privacy while preserving meaningful insights from the data.

Both Apple and Google collect data on usage statistics for malwares. However, they do it by maintaining differential privacy. Let's look at another technique called randomized response. Many times, surveys are on questions which are sensitive or stigmatized. For example, students may be asked whether they are using any substance, employees can be asked if they have taken bribes, etc. The respondents in these cases, will never give correct answers as they know if these responses are leaked, that can cause severe trouble to them. The objective of the surveyor will be defeated here.

So, let's say if the warden of a hostel wants to check on the habit of smoking among the students. The survey can be setup like this:

Do you Smoke?

Randomized Response: Flip a coin. If it is head, then tell the truth, else always say 'you smoke.'

Of course, what the respondent gets in the coin flip is not revealed or recorded. So, if the teacher wants to catch hold of a student and ask him why he or she smokes, he can definitely deny, telling that he/she got a tail, hence he or she said yes.

How does it help the surveyor in getting the percentage of smokers among the students? Well, if it is an unbiased coin, then the probability of head and tail are same i.e., 50%. So, let's assume x% students smoke. 50% of them will get head and will say they smoke, which will be $x/2$%. Rest, all people irrespective of they smoke or not are going to say, they smoke. Hence roughly another 50% will say they smoke. Hence, a total of $(50 + x/2)$ % are going to say, they smoke. After the survey, we will know the value of $(50 + x/2)$ % and consequently can easily find out what is x. There may be a subtle variation because we assumed exactly 50% heads and tails. However, this should be good enough at an aggregate level.

4.7 SECURITY OF THE AI MODELS

Security concerns regarding an AI model encompass a range of potential risks and vulnerabilities that could impact the confidentiality, integrity and availability of the model and its associated data. As earlier mentioned, not all the issues or concerns that we discuss are isolated. Data security issues will have some overlap with data privacy, transparency, etc.

Some key security considerations include:

- **Making the models safe against adversarial attacks**

  So, what is an adversarial attack?

  An adversarial attack in AI involves manipulating input data to deceive a model's predictions. Imagine you have an image recognition system that's great at identifying objects. An adversarial attack would involve making subtle changes to the input image, almost imperceptible to humans, but causing the AI model to misclassify the object. This type of attack reveals vulnerabilities in the model's decision boundaries and prompts efforts to enhance the model's robustness against such manipulations.

  Some of the strategies include adversarial training, gradient masking, ensuring input diversity, attack prediction, differential distillation and so on.

- **Making the models immune to deployment risks**

  Deployment typically happens in a different environment, which includes more software components besides the AI model. For access to the models, API (Application Programming Interface) may also need to be exposed. This environment and permissions need to be safeguarded against unauthorized usage and abuse.

  There is a quite famous experiment done by researchers from Carnegie Mellon University, where they showed that the AI-based facial recognition system fails when the person wears a particular type of patterned eyeglasses. These eyeglasses are what we call perturbation to the original pixels. In one of the cases, a white male's picture fooled the AI system into believing it was of a movie actress (Figure 4.3).

Figure 4.3 Adversarial attack on the images by just adding spectacle

4.8 INCLUSIVITY IN AI

In the context of AI, inclusivity refers to the design and development of artificial intelligence systems in a way that ensures fair and equitable treatment of diverse individuals and groups. It involves creating AI models and technologies that do not discriminate against or exclude people based on factors such as race, gender, age, ethnicity, disability, or other protected characteristics. Again, we would point out there is a decent amount of overlap between fairness and transparency. Hence, we would focus on aspects which plausibly, exclusively map to inclusivity.

- **Accessibility:** Designing AI applications to be accessible to people with diverse abilities and disabilities. This includes considerations for user interfaces, interaction methods and providing alternative means of access.

- **User-Centred Design:** Involving diverse stakeholders, including end-users from various backgrounds, in the design and testing phases to better understand different perspectives and needs.
- **Google Euphonia:** Google Euphonia is one such initiative. It aims to improve speech recognition for diverse populations. It utilizes a massive dataset of speech recordings from speakers with various accents, speaking styles, and disabilities. This data is then used to train AI models to better understand and transcribe speech, regardless of individual differences. This is really a very interesting initiative and interesting readers are welcome to browse more details at (https://sites.research.google/euphonia/)
- **SignAll:** This AI-powered app translates sign language into spoken language and vice versa. Again an excellent initiative (http://www.signall.live/)

4.9 SUSTAINABILITY IN AI

'Sustainability in AI' refers to the responsible and ethical development, deployment, and use of artificial intelligence technologies to ensure their long-term positive impact on society, the environment and economic systems. We can easily see, again there would be some component which overlaps with ethics and transparency. The most important factor is the environmental impact.

Environmental Impact

Addressing the environmental impact of AI technologies, including the energy consumption of AI models during training and inference. Sustainable AI encourages the development of energy-efficient algorithms and computing infrastructure. The training phase of large and complex neural networks, especially deep learning models, often requires significant computational power and resources. They demand powerful hardware, such as Graphics Processing Units (GPUs) or specialized hardware like TPUs (Tensor Processing Units). The computational intensity of training these models can result in substantial energy usage. They often are run from the data centres which have high energy requirements around the clock. The energy consumed by these AI computations is typically a traditional source of energy, hence increasing carbon footprint. Researchers and engineers are actively working on developing more energy-efficient algorithms and hardware architectures to reduce the environmental impact of AI. Techniques such as model compression, quantization and sparsity optimization aim to achieve similar or improved performance with reduced computational requirements. These are often called 'Green AI' practices.

It may be noted that AI can also be used for Sustainability.

As example, it can be used for optimizing resource management in smart grids or precision agriculture. AI can be used for actually most of the Sustainable Development Goals (SDG) as listed by the United Nations (https://sdgs.un.org/goals). If we focus on the first SDG that is eradication of poverty. AI has been used to (a) Identify poverty-prone regions; (b) When there are resources available to be distributed, identify the most needy person/household. Similarly, there are many such use cases for the rest of the SDGs.

'Distiller' refers to an open-source library for neural network compression and model optimization in PyTorch. Distiller is designed to facilitate the development and deployment of efficient deep learning models by providing a framework for model compression techniques such as pruning, quantization and knowledge distillation.

4.10 ROBUSTNESS AND RELIABILITY

Robustness in AI models refers to the ability of a model to maintain its performance and generalization capability under various conditions, including situations that differ from those encountered during training. A robust AI model should provide consistent and accurate predictions even when faced with noisy or unexpected input data, changes in the environment, or variations in the distribution of the data. Adversarial attacks, distribution shifts and outliers are common challenges to robustness. Robust models are less sensitive to minor perturbations in input data and can handle a diverse range of scenarios.

Reliability in AI models refers to the consistency and dependability of the model's predictions over time and across different instances. A reliable AI model produces consistent and trustworthy results. It should demonstrate stability and perform consistently across various inputs and under different conditions. Factors such as data quality, model architecture and training procedures can impact reliability. Reliable models are less prone to producing unexpected or erratic behaviour.

It may be of interest to users to note that, the models often face drifts primarily of two types namely concept drift and data drift.

4.10.1 Concept Drift

Concept drift occurs when the statistical properties of the target variable (the variable you're trying to predict) change over time. In other words, the relationship between the input features and the target variable evolves.

Example: If you have a model that predicts stock prices, concept drift may occur if the factors influencing stock prices change over time due to economic conditions, market trends, or other external factors.

4.10.2 Data Drift

Data drift refers to changes in the input features used by a model. It happens when the distribution of the input data changes, but the relationship between input features and the target variable remains constant.

Example: In a spam email classifier, data drift might occur if the characteristics of spam emails change over time, such as a shift in the language used or the types of phishing techniques employed.

4.11 CONCLUSION

AI through its different manifestations have become embedded in our life. For doing a task in lesser time or for a better-quality output we often take the help of AI-based models, algorithms, etc. One of the principal differences between AI-based algorithms with traditional computer science algorithms is these are not explicitly programmed and they learn on their own from the data. Hence, AI needs to be ethical and responsible. Some core issues related to AI ethics are data security and privacy, transparency, accountability, fairness, sustainability, robustness and reliability, etc. Each of these issues is discussed in detail with examples and recent developments.

4.12 SUMMARY

- Over the past two decades, AI programs have achieved significant milestones, from defeating human champions in games like chess and Go to impacting various sectors such as education, healthcare, transportation, and agriculture.

- As AI becomes integrated into daily activities, non-tangible issues like ethics, fairness and bias are emerging concerns, reflecting the need for responsible AI development.

- Ethics, in the context of AI, involves principles and standards guiding professionals in the Information Technology (IT) industry, ensuring morally and socially responsible conduct.

- Issues in AI development include the vast amount of data leading to potential biases, challenges related to data privacy and the lack of complete transparency in the internal workings of AI algorithms.

- The accountability of AI algorithms poses a dilemma, as they may produce unexpected results even with sound design principles, raising questions about transparency, interpretability and causality.

- Visibility into data and training, algorithmic transparency and understanding of decision-making processes are essential for transparency.

- The deployment of AI models, such as self-driving cars and judicial tools like COMPAS, raises ethical concerns, including decision-making in critical situations and potential biases in predicting recidivism.

- Ethics in AI involves principles such as fairness, transparency, accountability, privacy, security, inclusivity, sustainability and robustness, aiming to guide responsible AI development.

- Bias in AI systems is a critical concern, stemming from data bias, algorithmic bias and societal bias, with examples ranging from gender stereotypes in word embeddings to biased hiring practices in Amazon's AI-based recruitment tool.

- Bias detection involves analysing data, reviewing variations between groups and visually inspecting charts for unusual patterns. Fairness metrics, such as demographic parity, can be calculated to assess fairness in AI decision-making.

- Striking a balance is crucial to providing meaningful insights without compromising security, user comprehension, or a company's competitive edge.

- Concerns arise about data sharing's impact on privacy, as illustrated by the concept of Differential Privacy, which aims to protect individuals' privacy in aggregate studies.

- Inclusivity in AI involves fair and equitable treatment, addressing factors such as race, gender, age, and disability.

- Accessibility and user-centric design are essential, ensuring AI applications are usable by diverse populations.

- Sustainability in AI focuses on responsible and ethical development to ensure positive impacts on society, the environment and economic systems.

- Robust AI models maintain performance under various conditions and are less sensitive to perturbations. Reliability in AI models ensures consistency and dependability over time and across different instances.

4.13 PRACTICE EXERCISES

4.13.1 Subjective Questions (10 Marks)

1. The consideration of Ethics in AI Programs is different from traditional Programs. Justify the statement by enlisting difference between AI Program and traditional Programs.
2. Elaborate on the concept of bias in AI systems. Discuss the different types of bias, such as data bias, algorithmic bias, and societal bias, providing real-world examples for each.
3. Discuss the challenges associated with transparency and explainability in AI algorithms. Why is it important for AI models to be transparent, and what are the implications of a "black box" model?
4. Explain the concept of "differential privacy" and its significance in preserving privacy in AI systems. Provide examples and discuss the trade-offs involved in choosing different epsilon values.
5. Define and discuss the term "accountability" in the context of AI systems. How does accountability differ from liability and responsibility, and why is it crucial for ensuring ethical AI practices?
6. Discuss the significance of visual inspection in detecting bias in AI systems. Provide examples of how graphs and charts can be employed to identify potential bias.
7. Discuss on the overlapping concepts in Ethical AI, such as fairness, transparency, accountability, and privacy. Provide examples of how these concepts intersect.
8. How can fairness metrics contribute to addressing bias in AI systems, and what are their limitations?

4.13.2 Subjective Questions (5 Marks)

1. In the context of AI ethics, discuss the trade-off between the need for open datasets for research and the concerns about data privacy.
2. Explore the potential negative points associated with transparency in AI systems
3. Discuss the challenges and potential risks associated with the security of AI models.
4. Discuss the concept of k-anonymity as a solution and its application in preserving privacy in datasets.
5. Discuss the ethical implications of AI advancements in various sectors such as healthcare, transportation, and education.
6. Explain the process of detecting bias in AI systems using data review and visual inspection techniques. Provide examples to illustrate each step.
7. What are some of the guidelines to make the development and deployment of AI models sustainable.
8. How can AI models be made robust against adversarial attacks, and what measures should be taken to mitigate deployment risks?

4.13.3 Short Answer-type Questions (2 Marks)

1. What is an adversarial attack?
2. Explain what is societal bias?
3. What do you understand by differential privacy?
4. What is algorithmic transparency?
5. What is the primary focus of inclusivity in AI?
6. Define concept drift in context of AI models. Provide examples
7. Define data drift in context of AI models. Provide suitable example
8. What is Green AI ?
9. Explain the difference between robustness and reliability in AI models

4.13.4 Multiple Choice Questions

1. What is the key difference between traditional software and AI software in terms of learning?
   a. Traditional software requires explicit programming, while AI software is continuous and learns from data.
   b. Traditional software is data-driven, while AI software is rule-based.
   c. Traditional software adjusts to new situations, while AI software is inflexible.
   d. Traditional software applies to broader problems, while AI software is limited to specific tasks.

2. What is one of the non-tangible issues discussed regarding AI algorithms?
   a. Computational efficiency
   b. Fairness and bias
   c. Legal consequences
   d. Rule-based programming

3. What is one of the factors considered in achieving transparency in AI systems?
   a. Informed consent
   b. Computational efficiency
   c. Explainability and interpretability
   d. Societal bias

4. What is the key parameter in differential privacy that represents the level of privacy protection?
   a. Alpha (α)
   b. Beta (β)
   c. Delta (δ)
   d. Epsilon (ε)

5. In the context of AI models, what does robustness refer to?
   a. Consistency and dependability
   b. Ability to maintain performance under various conditions
   c. Ensuring ethical development
   d. Accessibility to diverse individuals

6. What does the term 'accountability' signify in the context of AI systems?
   a. Legal consequences
   b. Ethical and moral expectations
   c. Adherence to regulatory frameworks
   d. Anticipation and corrective measures for decisions and actions

7. What does the term 'responsibility' encompass in the context of AI systems?
   a. Legal consequences
   b. Ethical and moral expectations
   c. Adherence to regulatory frameworks
   d. Expectation of proper functioning throughout the AI system lifecycle

8. What does the term 'robustness' refer to in the context of AI models?
   a. The ability to maintain accurate and reliable performance across diverse conditions
   b. The transparency of the algorithm
   c. The continuous learning capability of the AI model
   d. The ethical considerations in AI development

9. What is 'Explainable AI (XAI)' focused on achieving?
   a. Enhancing model resilience
   b. Providing comprehensive documentation
   c. Making complex models more interpretable
   d. Improving transparency in user interface design

10. What should be an ideal first step in detecting bias in AI systems?
    a. Check fairness metrics
    b. Visual inspection
    c. Review the data
    d. Compare outcomes

11. What is a potential negative point associated with transparency in AI systems?
    a. Enhanced security risks
    b. Improved user comprehension
    c. Reduced competitive disadvantage
    d. Misinterpretation by users

12. How can visual inspection be utilized to detect bias in AI systems?
    a. By comparing fairness metrics
    b. By analysing the decision-making process
    c. By creating graphs and charts to identify the behaviour of the target variable across various groups
    d. By calculating transparency scores

4.14 ANSWER KEYS

4.14.1 Multiple-choice Questions

1. (a)   **2.** (b)   **3.** (c)   **4.** (d)   **5.** (b)   **6.** (d)   **7.** (b)   **8.** (a)
9. (c)   **10.** (c)   **11.** (a)   **12.** (c)

AI in Research, Generative AI and Other Important Issues

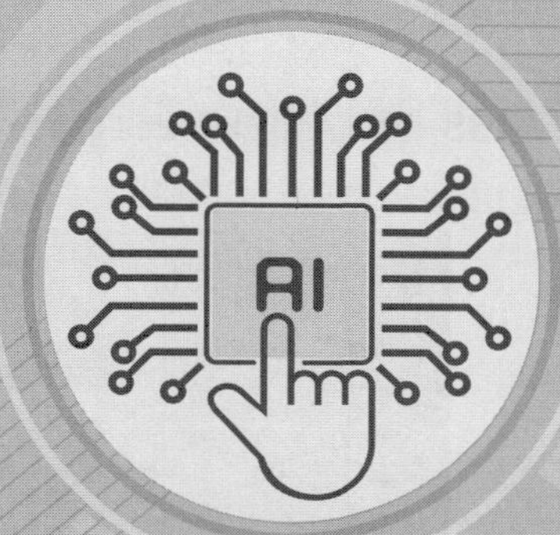

In the constantly evolving realm of scientific investigation, the integration of Artificial Intelligence (AI) stands out as a transformative influence, fundamentally altering the approach researchers take toward specific challenges. This chapter explores the convergence of AI and scientific research, delving into the dynamic contributions of intelligent algorithms as they contribute to the pursuit of knowledge. Additionally, the emergence of Generative AI has brought forth numerous possibilities and applications for AI. Within this chapter, there will be a broad introduction to Generative AI, accompanied by an overview of its most prominent product, namely ChatGPT. Notably, Prompt Engineering is highlighted as a crucial skill, poised to become the next significant expertise, as it plays a pivotal role in unlocking the true capabilities of ChatGPT. The narrative extends to include a discussion on other noteworthy advancements in the field of AI.

5.1 AI IN EXPERIMENTATION AND MULTI-DISCIPLINARY RESEARCH

There are many activities in experimentation and research where AI can be of help. Some indicative steps are the Generation of Hypothesis, Virtual Prototyping or Simulation, Collection of the data and then doing analysis or finding patterns. In the following section, we will explore the use of AI in various disciplines and sub-disciplines.

5.1.1 Application in Particle Physics (Identifying Particles in Collider Data)

At the Large Hadron Collider (LHC), AI algorithms are used to shift through massive amounts of data and accurately identify different particles, such as Higgs bosons, among billions of collisions. Located in Geneva, Switzerland this is definitely one of the wonders of Science and Engineering. Designed to collide protons and ions at close to the speed of light, recreating conditions similar to the Big Bang. This task would be extremely time-consuming for humans to do manually. Thousands of scientists from over 100 countries work on the LHC (Fig. 5.1).

CERN runs highly complex scientific instruments that require advanced control systems and produce enormous amounts of research data. Managing these complex systems and analysing massive datasets pose major challenges.

Figure 5.1 LHC Tunnel of CERN

To address these challenges, CERN is leveraging artificial intelligence techniques in various ways. AI is helping CERN improve particle beam handling through smarter control algorithms. It enables more efficient analysis of the petabyte-scale data generated by CERN's experiments. And AI techniques are being applied to optimize maintenance and upkeep of CERN's large-scale facilities. By harnessing artificial intelligence, CERN aims to enhance many aspects of operating its sophisticated scientific infrastructure for particle physics research.

5.1.2 Application in Astro Physics and Space

AI is used to analyse images of galaxies, stars, and other celestial objects to detect patterns, classify objects, and identify anomalies that might lead to new discoveries. Machine learning models can simulate the evolution of galaxies, predict the behaviours of stars, or simulate the effects of gravitational interactions. AI is used to simulate complex astrophysical phenomena, such as galaxy formation, star formation, and the dynamics of celestial bodies. This helps researchers understand the underlying processes and test theoretical models.

Scientists studying space have had a problem when creating simulations—they had to choose between making detailed simulations or covering a lot of the universe. Now, with the help of a technology called generative adversarial networks, they can do both.

Researchers from Carnegie Mellon University and the University of California came up with a smart computer program. This program can take basic simulations and turn them into highly detailed ones quickly. So now, scientists can make complex simulated universes in just one day. AI is also being used for mapping of dark matters (Fig. 5.2)

Figure 5.2 Mapping of Dark Matters using AI

5.1.3 AI in Experimental Chemistry

AI can assist in various aspects of experimental chemistry, offering benefits such as improved efficiency, data analysis and the discovery of novel compounds. The AI algorithms can analyse data from previous experiments to suggest optimal conditions for reactions, helping researchers design experiments with higher success rates. AI can predict reaction pathways and propose optimal sequences of reactions, facilitating the planning of complex synthe-ses. The AI algorithms can analyse spectroscopic data (e.g., NMR, IR, UV-Vis) to identify compounds, determine reaction progress, and characterize reaction products. It can help, process and analyse large datasets of chemical information, facilitating the extraction of meaningful patterns and insights. AI assists in the virtual screening of potential drug candi-dates, predicting their binding affinities to specific biological targets and prioritizing com-pounds for synthesis and testing.

Revolutionizing the art of molecule-making, AI retrosynthesis tools are rewriting the rules of chemical synthesis. These digital wizards take a target molecule as their canvas, then paint in reverse, meticulously charting the ideal pathway back to its simplest building blocks. One promising approach, 3N-MCTS, blends a timeless search algorithm with the power of three neural networks, unveiling the most efficient route from concept to creation. Despite the buzz these tools generate, the majority of chemists are yet to fully embrace their transform-ative potential.

5.1.4 AI in Biology

AI algorithms are deciphering the vast libraries of genetic information, identifying patterns and connections that would elude even the most dedicated human researchers. This is leading to breakthroughs in understanding diseases, personalized medicine, and even the evolution of life itself. Some of the use cases of AI in biology are as follows:

- Predicting protein structures from amino acid sequences using deep learning.
- Identifying disease risk factors and biomarkers from genetic data through machine learning.
- Automating the analysis of microscope images to detect cells, tissues, etc., via computer vision.
- Discovering new drugs and materials by screening virtual libraries with AI algorithms.
- Modelling complex biological systems like gene networks and cell interactions using Bayesian networks and machine learning.
- Synthesizing novel proteins with desired functions using generative AI models.
- Classifying bacteria species rapidly using image recognition and spectral data with neural networks.
- Tracking neural activity in the brain in real-time using AI signal processing and analytics.
- Mining scientific literature to extract biological knowledge and relationships using natural language processing.

One of the leading Research Organizations shows (Fig. 5.3) how Global AI Genomics is poised to grow.

Citation/AI Paper based on application Domain

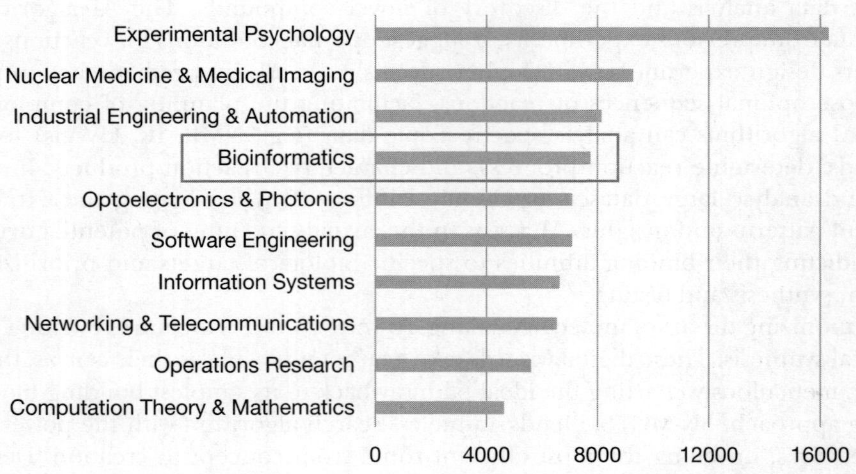

Figure 5.3 Growth Projection of Genomics Market in the US

5.1.5 AI in Environmental Science

Artificial Intelligence (AI) can be applied in various ways in environmental science to address challenges, improve efficiency and enhance decision-making.

- **Climate Modelling and Prediction:** AI can analyse vast amounts of climate data to improve climate models and enhance predictions. Machine learning algorithms can identify patterns, trends, and correlations in climate data, leading to more accurate and timely climate forecasts.

- **Air and Water Quality Monitoring:** AI technologies can be used to analyse sensor data and satellite imagery to monitor air and water quality. Machine learning algorithms can identify patterns and anomalies, providing insights into pollution sources, potential health risks, and environmental impacts.

- **Waste Management:** AI technologies, including robotics and machine learning, can be applied to improve waste sorting, recycling processes and waste management overall. Automated systems can identify recyclable materials and help reduce the environmental impact of waste disposal.

- **Resource Management and Conservation:** AI algorithms can optimize resource management strategies, such as precision agriculture. By analysing data on soil conditions, weather patterns and crop health, AI can help farmers make more informed decisions, reduce resource use, and minimize environmental impact.

POINTS TO PONDER

AI for Penguins

In 2017, there was a very interesting competition launched by DrivenData. Antarctic penguins are hugely popular and appealing animals known for their distinctive behaviour. Their quirky nature has been documented extensively by wildlife photographers and researchers. To better manage and conserve penguin species over the long term, accurately modelling and predicting their population numbers is essential. It also provides insights into the dynamics of the broader Antarctic ecosystem. This data science competition aimed to leverage the skills of the data science community to develop models that can effectively forecast penguin population changes. By tapping into innovative modelling approaches, the goal was to produce more accurate and robust predictions of penguin populations and their shifts. Reliable models for anticipating population trends are vital tools for enacting evidence-based conservation policies for Antarctic penguin species. One of the prominent species is Emperor Penguin (Fig. 5.4).

Figure 5.4 Herd of Emperor Penguin

5.2 GENERATIVE AI INTRODUCTION

If we look at this topic from just a semantics perspective, generative AI is a specialized branch of AI, which can generate new content. That content may be new text, that may be images, videos, songs and of course a mixture of the same. Specifying that it can generate is not enough. The content should be such that, it is coherent and must pass the 'Turing Test', i.e., it should not appear as machine-generated. Predominantly, deep learning or neural network-based models are used as generative models.

Machine learning or deep learning models can be broadly classified into (a) Discriminative Models and (b) Generative Models. Enlisted are some of the key points about these models.

1. **Discriminative Model:**
   - Models' conditional probability $p(y|x)$ of outputs y given inputs x
   - Focuses on predicting y from x, not the full input distribution
   - **Examples:** Logistic Regression, SVMs, Neural Networks
   - Does not generate new data points
   - Better for classification and prediction tasks

2. **Generative Model:**
   - These models the joint probability $p(x, y)$ of inputs x and outputs y
   - Learns the distribution of all input data
   - **Examples:** Naive Bayes, LDA, Gaussian Mixture Models
   - Can generate or sample new data points
   - Good for modelling input variability

In a layman's term, given different attributes of an unknown species, discriminative model will have the capability to label it to one of the known categories. It will say this looks more like a dog, than a cat. The generative model will try to understand what makes a dog a dog. Try to unearth the underlying probability distribution behind the same. As a result, it will develop the capability of generating new dogs. I think, it is not difficult to understand that the goal of the generative model is much more lofty.

The current generative models are backed by a huge amount of data. They are gigantic models with billions of parameters. Many big organizations were working on building generative models.

5.2.1 Chronology of the Different Developments

However, there is no denying that OpenAI was the first to take the world by storm and showed what is the capability of a generative AI model in a democratized fashion. Soon other giants like Meta and Google joined the bandwagon. There have been a series of developments both in case of text and image generation, which has led to the current state of generative AI. In the, following section, we quickly do a time travel on these inventions.

- **1960-70:** n-gram models and Markov models were being used for language modelling. They continue to advance in the areas of Natural Language Processing for the subsequent two decades.

- **1980:** Autoencoders, though they were introduced for unsupervised learning, they provided a foundation for the encoding and decoding architecture of generative models. It is to be noted, much of autoencoder's glory was achieved later.
- **2003:** Yoshua Bengio and team created the first feed-forward neural network language model, capable of predicting the next word in a sequence. Predicting the next word is indeed a small but certain step towards generative AI
- **2014:** Generative Adversarial Networks (GANs) are introduced by Ian Goodfellow, revolutionizing the field by pitting two neural networks against each other, one generating content and the other discriminating between real and fake, leading to increasingly realistic outputs.
- **2014:** Variational Autoencoders - Deep latent variable models for generating new data points. The vanilla autoencoders have a deterministic behaviour, whereas the Variational Autoencoder brings an element of chance or probability, making their capability of generation, far more enriched.
- **2016:** PixelRNN/PixelCNN - Models for sequentially generating images pixel-by-pixel.
- **2020:** OpenAI unveils GPT-3, a large language model that exhibits remarkable fluency and coherence in text generation, sparking discussions about the potential and limitations of such powerful AI systems.
- **2022:** DALL-E 2, a diffusion model developed by OpenAI, takes image generation to new heights, producing stunningly realistic and creative visuals based on text prompts. Noise introduction as well as elimination is done in step-by-step manner in a diffusion model to generate a realistic image. The theory of Markov Chain is leveraged in the process.

Figure 5.5 is an Image created using Generative AI.

Figure 5.5 Image created by Generative AI

DO YOU KNOW

A deepfake is a synthetic media created using artificial intelligence (AI) and machine learning techniques to manipulate existing video or audio footage. This manipulation can range from subtle changes like altering someone's facial expressions to creating entirely new and realistic content featuring people who never actually said or did what is shown. In March 2023, a Canadian couple fell victim to a deepfake call scam, losing $21,000. The scam involved a fraudulent call where the perpetrator utilized technology to mimic the voice of the couple's son. The imposter falsely claimed to be in legal trouble, urgently seeking financial assistance for bail.

5.2.2 Synopsis of the Important Models

N-gram Models

N-gram models analyse sequences of words (n words at a time) to guess what's likely to come next. Like detectives gathering clues, they count how often words appear together, building a statistical map of language patterns. Unigrams consider single words, bigrams focus on pairs and trigrams analyse three-word sequences. The bigger the n, the more context they use, but also the trickier it gets to find enough examples in real text. Though simpler models, n-grams played a crucial role in early language processing and laid the groundwork for more powerful AI approaches.

- **Unigram:** Predicting the next word, without the context.
- **Bi-Gram:** Predicting the next word based on the last word
- **Tri-gram:** Predicting the next word, based on the last two words

Figure 5.6 depicts a lot of things.

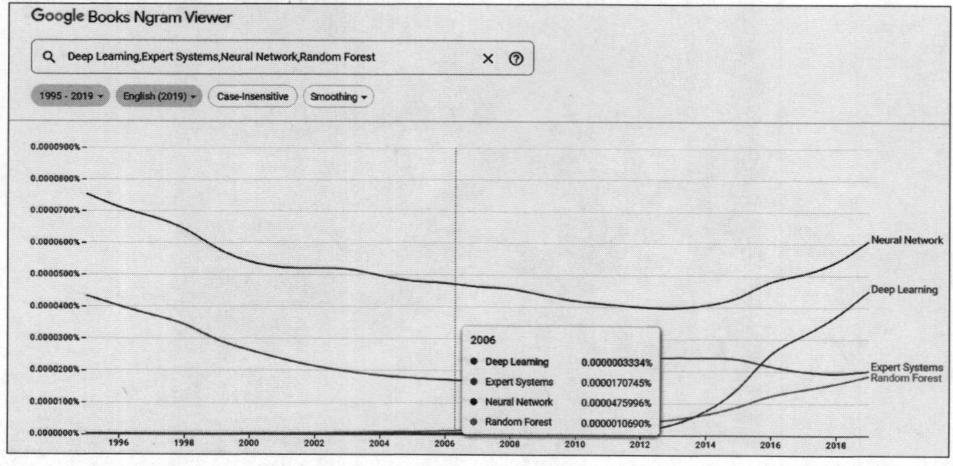

Figure 5.6 Google Books N-gram view of Deep Learning, Neural Network, Random Forest and Expert Systems

Google Books Ngram view allows us to see how these probabilities changed over the years. We find the cross-over pattern between deep learning (Part of Inductive AI) and Expert Systems (Part of deductive AI) really fascinating and revealing.

Auto-Encoders

- Autoencoders are a type of artificial neural network used for unsupervised learning and dimensionality reduction.
- They compress the input data into a lower-dimensional latent space representation and then reconstruct the output from this representation. For example, input the numbers 2, 6, 10, 14, 18, 22, 26 and the output is 30. So, there are 7 input variables. However, if we observe closely, the inputs are actually coming from an arithmetic progression, 2 is the first term and 4 is the common difference. Hence, rather than using the 7 inputs, we can use three inputs which are the first term, common difference and position of the term in the sequence.

$$F\ (2, 6, 10, 14, 18, 22, 26) \rightarrow 30$$

$$G\ (2,4,8) \rightarrow 28$$

As you see F is a function which works on the original input and G works on the first term, common difference and position of the term. These three numbers combined can be thought of as the latent representation. Now, neither in every problem, it is possible to find the latent representation, nor in those cases the latent representation will be easily comprehensible as shown in this example.

- An autoencoder consists of an encoder that maps the input to a latent code, and a decoder that reconstructs the input from the code (Fig. 5.7).

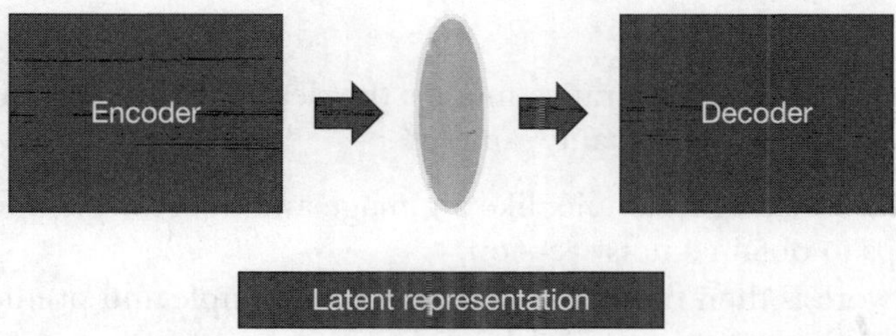

Figure 5.7 Block Diagram of Auto-Encoder

- The aim is to minimize the difference between the input and reconstructed output by optimizing the model parameters.
- Now, this latent space can be sampled and inputs can be given to the decoder, which will generate new points.

Generative Adversarial Networks (GANs)

The working of GANs is really interesting.

- Imagine two artistic rivals, constantly pushing each other to new heights. That's the essence of Generative Adversarial Networks (GANs)

- ▪ **Generator:** An artist, trying to create realistic paintings (e.g., portraits, landscapes).
- ▪ **Discriminator:** A critic, trying to spot fake paintings among real ones.
- The generator keeps improving its forgery skills, generating ever-more convincing artworks.
- The discriminator sharpens its detective skills, getting better at identifying fakes.

DID YOU KNOW

There is an incredibly cool thing called as Style Transfer. Let's say, we have a regular photo of a bustling city street and we want to render it Van Gogh's iconic swirling, streaky brushstroke style. This is totally possible using a popular GAN model named as StyleGAN. StyleGAN is a novel generative adversarial network (GAN) architecture for high-quality image generation developed by NVIDIA researchers in 2018–2019. It incorporates ideas from style transfer to give more control over image synthesis compared to regular GANs. It provides fine-grained control over image generation through injections of style vectors. The generator comprises a mapping network and synthesis network to disentangle style and noise. The mapping network transforms the latent code into an intermediate style representation. The synthesis network then generates the image from style and noise.

Next important set of models are the transformer models and the neural-based embedding models for text generation. But we postpone the discussion to the next section, where we discuss about ChatGPT and Prompt Engineering.

Diffusion Model

Diffusion models are a class of generative models that learn to reverse a multistep diffusion process in order to generate high-quality samples.

- They start with a real data sample, like an image and inject noise into it over multiple diffusion steps to obtain a noisy version.
- A neural network is then trained to take this noisy sample and predict how to remove some noise and reverse one step of the diffusion.
- By repeating this denoising process over many iterations, the model is able to remove all the injected noise and produce a realistic generated sample.
- Diffusion models are trained by optimizing the noise removal reconstruction at each step to mimic the data distribution.
- Key benefits are the ability to generate sharp, high-resolution outputs and excellent sample diversity.
- Recent variants like DDPM and GLIDE have shown impressive image and audio generation capabilities.
- The gradual denoising helps avoid spikes and artefacts common in other generative models.
- Diffusion models offer a noise-based perspective to generative modelling complementary to GANs and autoregressive models.

In summary, diffusion models take a stochastic approach to generation by training neural networks to model incremental noise removal from data. This yields highly realistic and diverse outputs across modalities.

5.3 CHATGPT AND PROMPT ENGINEERING

5.3.1 ChatGPT Context Setting

Let's quickly recap the development happening in NLP. There is no denying that machine-to-machine translation is one of the most complex tasks. This is a sequence-to-sequence task. i.e., input is a sequence and output is also a sequence. Regular neural network fails on sequence tasks. RNN and LSTM models came to the rescue. Here, the output is a function of the previous step output as well as the current state input.

However, in the real world when an interpreter tries to convert a sentence from one language to the other, he probably first listens to the entire sentence, understands the meaning and then states that in another language. So, it's not a word-by-word translation, rather arriving at an understanding and then restating it in the target language. Well, this is the intuition behind the encoder decoder network. As expected, this gave much better result than that of RNN and LSTM.

The next revolution in NLP came in terms of transformers. They used the encoder, decoder block and augmented it with attention. From an intuition perspective, this was more flexible than encoder, decoder. It's like we are not having the step of understanding the sentence fully and then regenerating, rather the transformer can look at the sentence in the source language in its entirety in the same instance while generating a word in the target language and it pays attention to only the source language words that are relevant. One other advantage of transformers over encoder decoder is they can be trained in parallel.

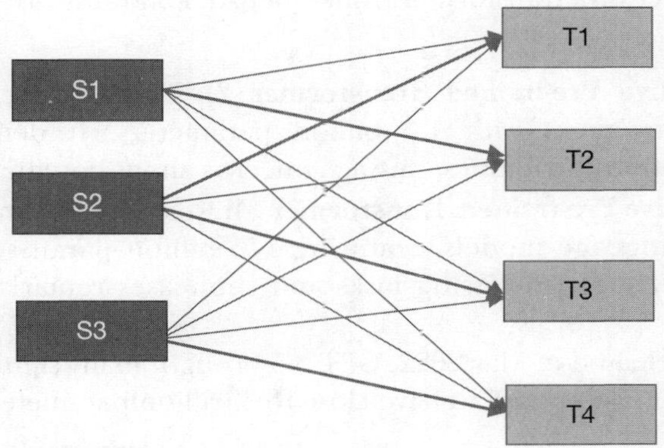

Figure 5.8 Attention Mechanism in Machine Translation

In Fig. 5.8, S1, S2 and S3 denote words in a sentence of the source language and T1, T2, T3 and T4 denote the words in the target language. There is all possible interconnection between each source and target pair of words. The thick lines indicate that while generating T4, attention needs to be made especially to S3. Such an interconnection layer, which allows to give specific weight to the inputs is called an attention layer.

In a more generic extension, there can be multiple layers of attention. So, one layer may look at the actual words, one layer can look at the part of the speech of the word. There can be several ways of paying attention to a word. One may be looking at the syntax of the words, another may be looking at the semantics of the words, yet another may be looking at the tone of the words. This technique is called multi-head attention.

With the advent of transformers, there was indeed a revolution in the way NLP tasks were performed earlier. While some transformers retained the encoder-decoder architecture, some altered these basic blocks.

The models that emerged as leaders initially were BERT (Bi-directional encoder representation) from Google and GPT (Generative Pre-Trained model) from Open AI. BERT's focus is more on understanding of language. As the name suggests, this focuses on the encoder part. Whereas GPT had its focus on the generative side i.e., the decoding part.

Apart from attention and parallel training, one other important characteristic of these transformers is pre-training. In terms of the number of parameters, amount of data, or CPU power these are gigantic beasts. Essentially, they are at first trained on large amounts of text mostly publicly available from Wikipedia or similar sources, even before they are given any language tasks to perform.

Before we move to elaborate on how the transformers are playing an important role, you may want to revisit, the important text processing tasks, that were discussed in Chapter 2.

Let's head to ChatGPT Now.

5.3.2 ChatGPT

ChatGPT is a super successful transformer from OpenAI. Following are the different versions of GPT.

- **GPT-2 (Generative Pre-trained Transformer 2):** Released in 2019, GPT-2 was a significant advancement with 1.5 billion parameters. It demonstrated powerful language generation capabilities, raising concerns about potential misuse.
- **GPT-3 (Generative Pre-trained Transformer 3):** Released in June 2020, GPT-3 is one of the largest language models, featuring 175 billion parameters. It excels in various natural language processing tasks and showcases remarkable text generation abilities.
- **ChatGPT 3.5:** Released in May 2022, GPT 3.5 brought many significant improvements and now GPT 4.0 is supposed to have close to 3 trillion parameters.

Let's quickly explore, some of the basic features of ChatGPT. The ensuing discussion is done in a generalized manner, so this will hold for Google BARD or Claude or any other similar tool.

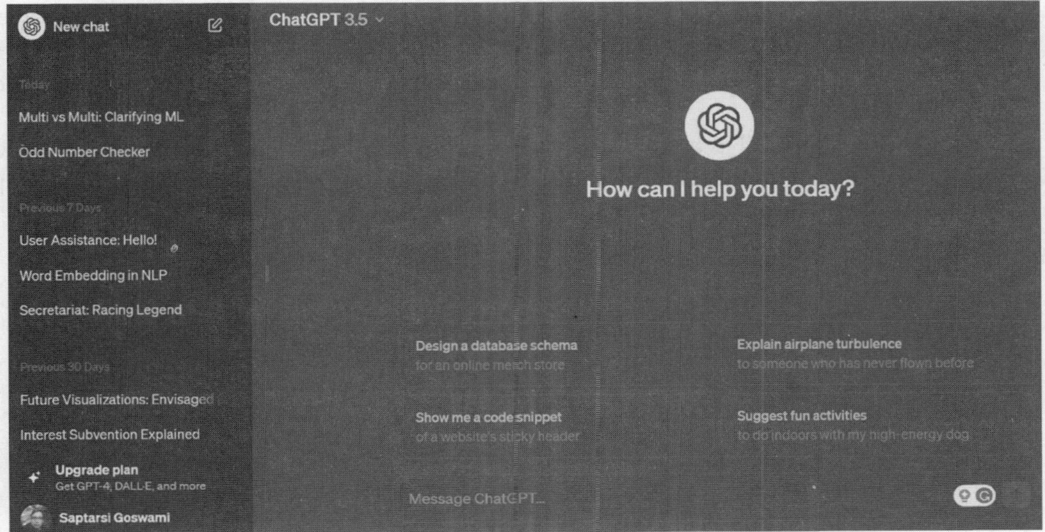

Figure 5.9 OpenAI ChatGPT Home Screen

If we look closely at Fig. 5.9:

- In the message part, you can ask a question and the question histories are saved.
- The question you ask is called a prompt. In the next section, we are going to read more about prompts (Fig. 5.10).
- These are probabilistic models; hence the result may be different to even the same question asked
- There will be an option to edit the question and resubmit.
- BARD can take input as audio as well as Images

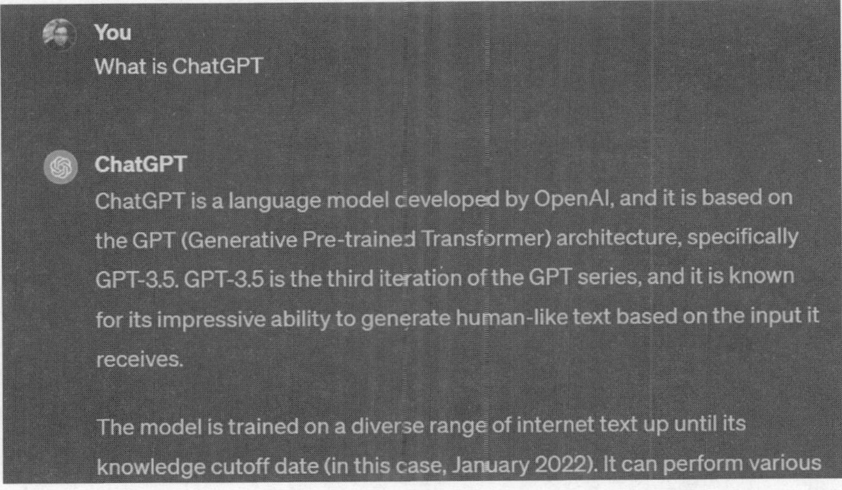

Figure 5.10 Prompt to ChatGPT and the response

- ChatGPT supports 90 + languages. If you wish to get output in a particular language, you will have to give the input also in that.
- After the response is generated, there are some options for liking, disliking and regenerating.
- Here's how, ChatGPT is continuously learning from humans, immediately it can know the feedback. In some cases, if you say, you dislike the response, it will go one step ahead and give another response and ask for your feedback again.

Below is a screenshot (Fig. 5.11) of when BARD was asked a question in Bengali.

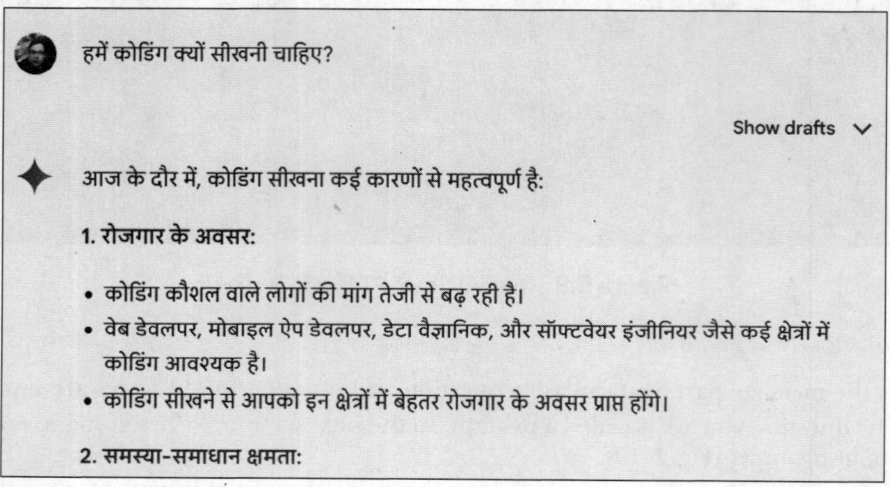

Figure 5.11 BARD now Gemini answering in Bengali

In the below instance, Google BARD can be found to be actually describing a diagram (Fig. 5.12)

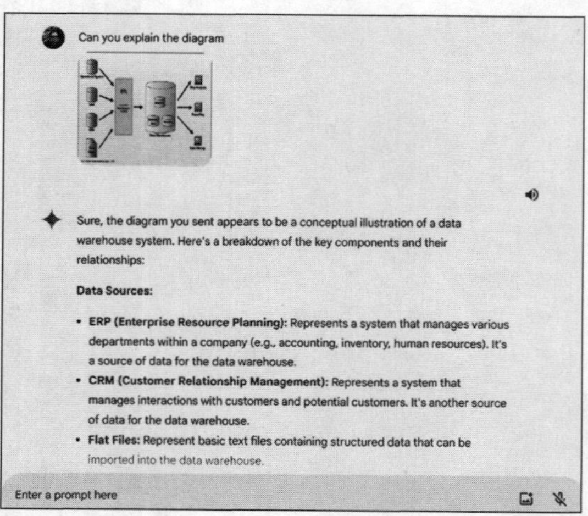

Figure 5.12 BARD now Gemini describing a diagram

5.3.3 Prompt Engineering

The questions that we ask to ChatGPT are called prompts. While ChatGPT can do exceedingly well, in answering it also needs well-formed questions, which are clearly formulated has the context associated with it. This art of asking the right question or formulating the right prompt is called prompt engineering. It is being touted.

This can be very clearly observed from the following Google Trends Graph (Fig. 5.13).

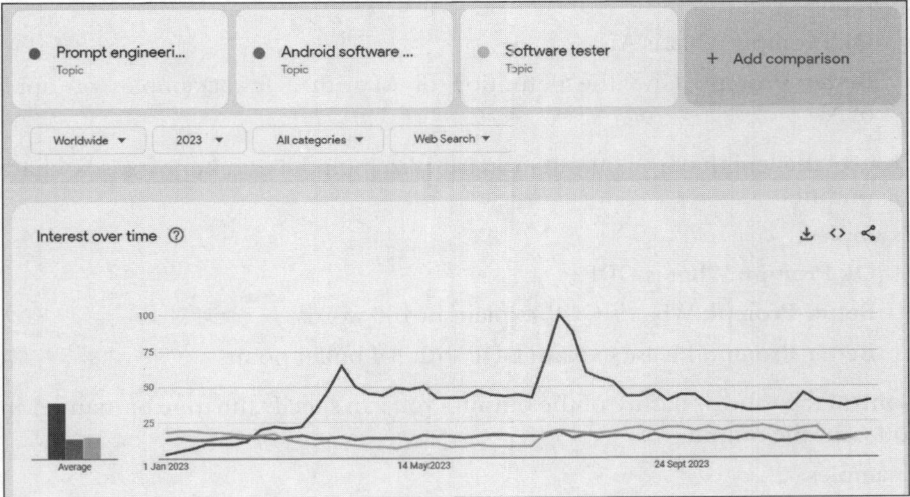

Figure 5.13 Comparing Prompt Engineering, Android, Software Tester over time using Google Trends

Prompt Engineering is searched more than, traditional software skills like Android Development or Software Testing.

In the below diagram, we can see, the affinity of different countries to prompt engineering in the world map (Fig. 5.14).

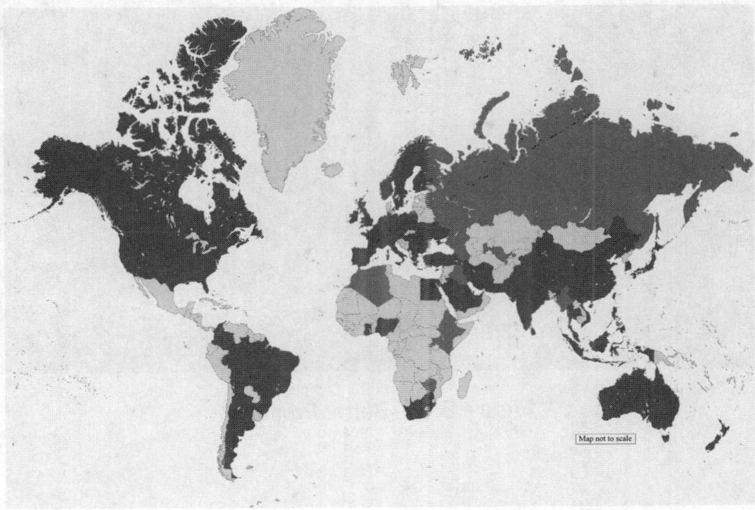

Figure 5.14 Comparing Prompt Engineering, Android, Software Tester over time using Google Trends

As it can be observed the entire world is curious about prompt engineering (Regions marked in darker shade), except areas like Russia and some low volume search regions (Which are greyed out).

The key aspects of prompt engineering include:

- **Task Framing:** Expressing the desired task clearly in natural language at the start of the prompt.

  Examples:

  - **Ok Prompt:** What is AI?

  - **Better Prompt:** Give the definition of AI with a few examples on application of AI

- Control the length of the output - You can also limit the number of words, character in the output.

  Examples:

  - **Ok Prompt:** What is GDP?

  - **Better Prompt:** What is GDP, explain in 100 words.

  - **Better Prompt:** Please explain GDP with 3-4 bullet points

- Control the tone or nature of the output - You can specify the tone or manner in which you want the output.

  Examples:

  - **Ok Prompt:** What is Industry 4.0?

  - **Better Prompt (Fig. 5.15):**

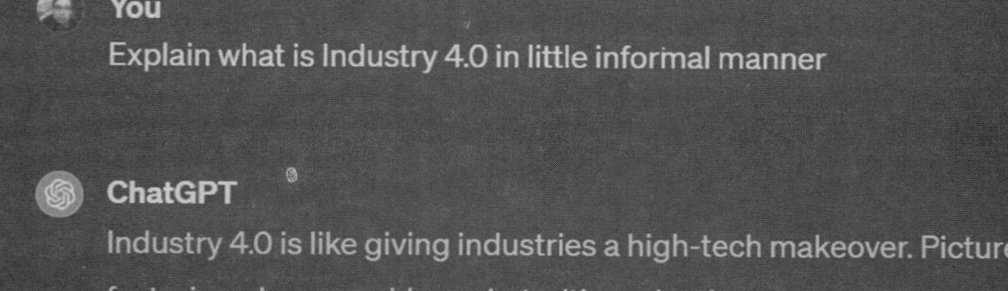

Figure 5.15 Better Prompts

▪ **Better Prompt (Fig. 5.16):**

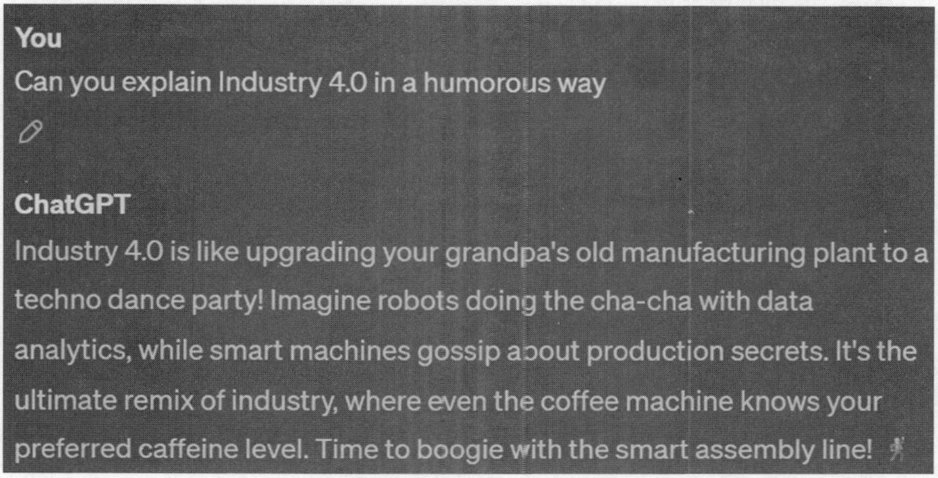

> **You**
>
> Can you explain Industry 4.0 in a humorous way
>
> **ChatGPT**
>
> Industry 4.0 is like upgrading your grandpa's old manufacturing plant to a techno dance party! Imagine robots doing the cha-cha with data analytics, while smart machines gossip about production secrets. It's the ultimate remix of industry, where even the coffee machine knows your preferred caffeine level. Time to boogie with the smart assembly line!

Figure 5.16 Better Prompts

- **Examples:** Providing a few demonstration examples that illustrate the input and expected output. This can be seen in the below diagram (Fig. 5.17).

> Give me list of 10 popular books for children.
> The sample output format will be like this
>
> 1. **Harry Potter and the Philosopher's Stone** by J.K. Rowling, published by Bloomsbury in 1997
> 2. **The Cat in the Hat** by Dr. Seuss, published by Random House in 1957
> 3. **Matilda** by Roald Dahl, published by Jonathan Cape in 1988
> 4. **Charlotte's Web** by E.B. White, published by Harper & Brothers in 1952
> 5. **Where the Wild Things Are** by Maurice Sendak, published by Harper & Row in 1963
> 6. **Alice's Adventures in Wonderland** by Lewis Carroll, published by Macmillan in 1865
> 7. **The Chronicles of Narnia: The Lion, the Witch and the Wardrobe** by C.S. Lewis, published by Geoffrey Bles in 1950
> 8. **Winnie-the-Pooh** by A.A. Milne, published by Methuen & Co. Ltd. in 1926
> 9. **The Gruffalo** by Julia Donaldson, published by Macmillan in 1999
> 10. **Green Eggs and Ham** by Dr. Seuss, published by Random House in

Figure 5.17 Prompt with specific output format

- One excellent technique here is to 'Explain to a 4th Grader' technique (Fig. 5.18). If there is a complicated concept, which you want it to be stated in a simple manner comprehensible to a 10-year-old, this can be used.

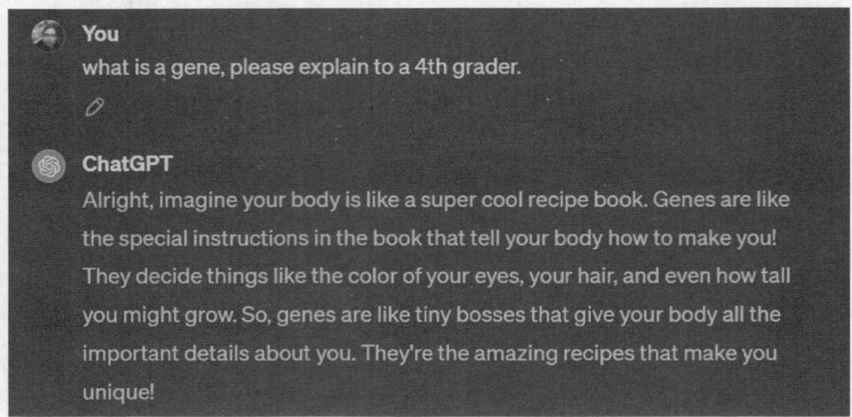

Figure 5.18 Fig.5.16 Explain to 4th Grader Technique

- Another useful prompt technique is role play. You can ask to role play as a fitness coach, a nutrition expert, a motivational speaker and so forth (Fig. 5.19).

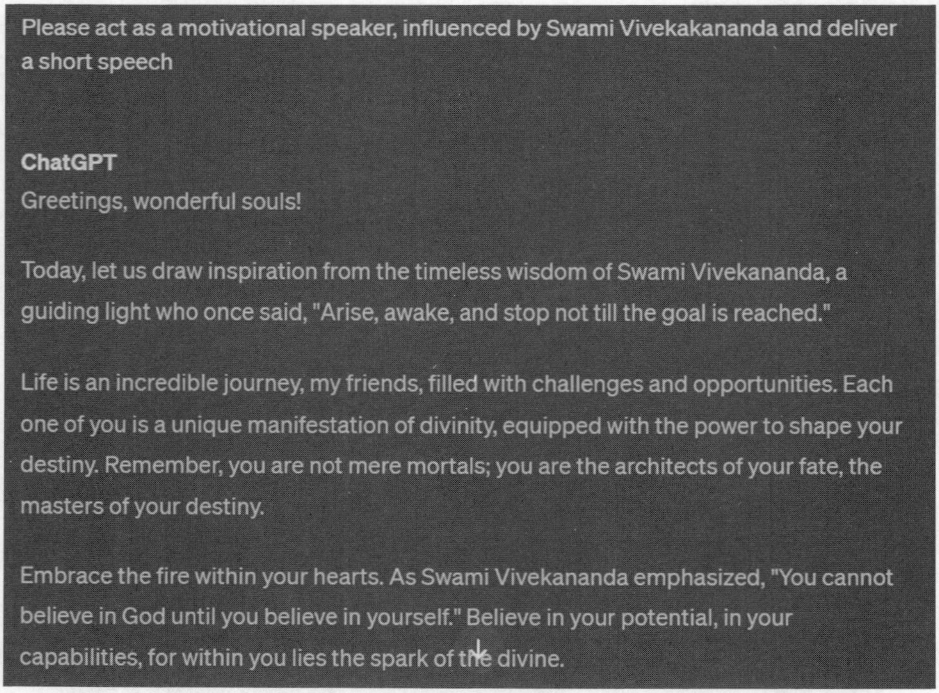

Figure 5.19 Prompt Engineering – Role Play or 'Act as a'

You can be really imaginative and creative in designing the prompts, which cater to your particular needs.

- **Prompt Priming:** Prompt Priming is a technique, where you take a very step-by-step approach. Rather than a direct question. Let's say, we want to get a delicious chicken recipe, rather than asking it directly, we can start by saying 'We are having some guests today, who are Punjabi, what kind of food will be good for them'. The very first option, that it gives is:

  'Butter Chicken (Murgh Makhani): A creamy and indulgent tomato-based curry with tender chicken pieces.'

  Then, you can say about other dishes, like you are going to eat with roti or rice, if you have already some gravy items and then you may specify that.

- **Chain of Thoughts (CoT):** In 2022, researchers at Google including Wei proposed an approach called Chain-of-Thought (CoT) prompting to improve large language models' performance on certain complex tasks. This method has the AI break down its thought process into intermediate reasoning steps.

  Below is an example of CoT for solving a math word problem:

  Demonstration: "A train travels 100 km in 2 hours. What is its speed?"

  Answer: '50 km/h. (Explanation: distance/time = 100 km/2 h = 50 km/h).'

  Prompt: 'A car travels 150 km in 3 hours. What is its speed?' Follow the steps from the demonstration to solve this problem.

- **Few Shots Learning:** In prompt engineering, few-shot learning is a technique where we leverage the capabilities of large language models (LLMs) to perform specific tasks by providing them with a limited number of examples (typically 2-10). It essentially teaches the LLM by demonstration, helping it grasp the desired output format and style without extensive training on large datasets.

  - Many of the tasks, ChatGPT or other agents can do, without any example. This is called as Zero-Shot learning.
  - When one example is given, it is often called One-Shot Learning

While the above discussion is focused on text, the below table gives a synopsis of tasks related to other type of contents and the tools associated.

Type of Content	Type of Task	Tool
Image	Image Generation from Text	Midjourney, Dall-E 2
Audio	Music Generation	MuseNet
Video	Generate Facial Expressions	FaceSwap
Audio	Voice Cloning	Resemble
Image	Photorealistic Images	Artbreeder

Following point regarding the examples and tools need to be noted

- This table and the examples using ChatGPT/ BARD (Gemini) is not exhaustive and represents some of the popular tools available.
- Tool availability and capabilities may change over time.
- Some tools may have limitations or require paid subscriptions.

5.4 EMERGING TRENDS AND FUTURE DIRECTIONS IN AI

Some of the important trends and directions which are going to bring disruptive and transformative changes are as follows:

Technological Advances

- **Advances in Deep Learning:** Such techniques include transformers, diffusion models, and foundation models which are essentially larger models with billions of parameters, trained on massive datasets. They are capable of multimodal AI that leverages vision, language and robotics. Increased integration of modalities.
- **Distributed Systems and Federated Learning:** Federated learning is an approach to machine learning where the model is trained across multiple decentralized devices or servers holding local data samples. Instead of sending raw data to a central server for training, the model is sent to individual devices, which compute updates based on their local data. These updates are then aggregated to improve the global model.
- **Causal AI and Machine Reasoning:** Causal AI is a subfield of AI that focuses on understanding and reasoning about cause-and-effect relationships. Unlike traditional AI which excels at prediction based on patterns, Causal AI aims to explain why things happen and how interventions can influence outcomes. AI chips and specialized hardware. New processors optimized for AI workloads.
- **AI Chips and Specialized Hardware:** GPUs have become a standard for training neural networks given their parallel processing power. Nvidia GPUs are widely used for deep learning. Competition is emerging from dedicated AI accelerators like TPUs (Tensor Processing Units) developed by Google specifically for machine learning workloads.

Responsible AI Practices

- With AI being an intertwined part of everyday life, there is an increased focus on AI safety, ethics and alignment. Hence, the key themes looking forward are increasing scale and computing, safety and ethics, making AI accessible and beneficial, and gaining a deeper understanding of reasoning, causality and the world. This we have previously discussed in Chapter 4.

AI with its Expanded Reach

- Some other trends are personalization and customization of AI where the models are tailored for individual users and use cases. AI is being used increasingly in science, automating and augmenting scientific discovery in materials, medicine, climate etc. There is emphasis on democratization of AI by easier access through pre-trained models, libraries, tools, etc.

DO YOU KNOW

CRFM, a Stanford HAI initiative, brings together faculty, students and researchers from over 10 departments to advance the science and responsible development of foundational language models. With a focus on ethics and societal impact, CRFM at Stanford HAI is dedicated to ensuring AI's powerful language models are used responsibly and contribute positively to society. The more can be found at https://crfm.stanford.edu/

5.5 AI AND SOCIAL INEQUALITY

AI in one hand can increase inequality again it has also the potential of mitigating the risk.

AI and Social Equality

AI can potentially automate many tasks. Automation may displace jobs, which may hit the lower income roles hardest.

Loss of Jobs

- **Routine and Repetitive Tasks:** Jobs involving repetitive manual tasks, such as assembly line work or data entry, may see a decrease in demand as AI and robotic automation become more advanced.
- **Customer Service:** Basic customer service roles, especially those involving routine inquiries or transactions, may be automated using AI-powered chatbots or virtual assistants.
- **Data Analysis:** Jobs that primarily involve data analysis and reporting could be affected, as AI algorithms can process and analyse large datasets more quickly and efficiently than humans.
- **Driving and Transportation:** With advancements in autonomous vehicles, jobs related to driving, such as taxi drivers and truck drivers, may see changes. Automated delivery services could also impact certain logistics jobs.
- **Manufacturing and Assembly:** AI and robotics can be used for tasks like manufacturing and assembly, potentially reducing the demand for manual labour in certain industries.
- **Routine Healthcare Tasks:** Jobs involving routine healthcare tasks, such as medical data entry and some diagnostic processes, could be affected as AI systems are developed for these functions.
- **Financial and Accounting Tasks:** Some financial and accounting tasks, including basic bookkeeping and data entry, could be automated using AI systems.

AI tools and services if are available to only a privileged few, that can create a divide between these groups. Use of AI needs access to computing devices, internet and sometimes subscription to services. These facilities are not typically available to all. Those with limited access to education and resources may lack the skills needed to engage with and benefit from AI technologies, creating a knowledge as well as skills divide. Some other issues like Bias, lack of diversity can produce results which are not best for equality. These issues are also elaborated in detail.

How AI Could Promote Equality

AI technologies can improve access to education, healthcare, and information, bridging gaps and providing opportunities to underserved communities. There are many AI solutions for children with special needs.

Khan Academy Kids, this free platform utilizes an AI-powered learning engine that personalizes the learning experience for each child. Language-learning apps like Duolingo use AI to give feedback and support to new language learners from underserved communities. Inclusive design, testing and auditing of AI systems by diverse teams improve social awareness.

5.6 SUMMARY

- AI aids in generating hypotheses, virtual prototyping, data collection, and analysis in various scientific activities.

- AI finds application in Particle Physics, Astro Physics and Space Experimental Chemistry, Biology, Climate Modelling and many more

- Generative AI is a specialized branch of creating new content like text, images, video and music.

- **Timeline of developments:** N-gram models, autoencoders, GANs, Variational Autoencoders, PixelRNN/PixelCNN and recent breakthroughs like GPT-3 and DALL-E 2.

- N-gram Models analyse word sequences, with unigrams, bigrams, and trigrams predicting the next word based on context. Autoencoders are neural networks for unsupervised learning and dimensionality reduction, compressing input data into a lower-dimensional latent space. Generative Adversarial Networks (GANs) use Artist (Generator) *vs.* Critic (Discriminator) concept, creating ever-more convincing artworks.

- Diffusion Model is a class of generative models for high-quality sample generation. Injects noise into real data over multiple steps, denoising to generate realistic samples.

- Transformers use attention mechanisms and parallel training for more flexible and efficient processing. ChatGPT introduced by OpenAI is based on transformer technology with outstanding capability in standard NLP tasks include spam detection, NER recognition, sentiment analysis, machine translation, text summarization, Q&A, and text generation.

- Prompt Engineering is a technique of crafting well-formed and contextually clear questions for ChatGPT, known as prompts.

- Some of the techniques to improve the prompts are task framing, controlling output length and tone, Prompt Priming, 'Act as', 'Explain to', etc.

- Advanced Prompt Engineering Techniques include Chain of Thoughts where LLMs are guided to provide a step-by-step reasoning for solving a mathematical or logical problem.

- Few Shots learning, on the other hand, provides the LLMs with some examples, which are to be followed in producing the output.

- Some of the technology advances of course include technology advances in deep learning, federated learning, specialized hardware like GPU, TPU. AI is getting democratized and is available for everyone's use. However, with the widespread use of AI, there is an increasing thrust on responsible AI.

- AI posed challenges like loss of Jobs, Access Divide, Bias and Lack of Diversity, which influence decisions. Some of the opportunites it presents are improved access and inclusive design.

5.7 PRACTICE EXERCISES

5.7.1 Subjective Question (10 Marks)

1. How does AI contribute to experimentation in various scientific disciplines and what specific activities can AI assist in?

2. How can AI be applied in environmental science, specifically in climate modelling, air and water quality monitoring, waste management and resource conservation?

3. Define Generative AI and its role in generating new content, highlighting the importance of coherence and passing the Turing Test.

4. Differentiate between discriminative models and generative models in machine learning, providing examples of each.

5. Outline the chronology of developments in generative AI, emphasizing key milestones and breakthroughs from 1960 to 2022.

6. What are the significant contributions of OpenAI in the field of generative AI and how have other major organizations, such as Meta and Google, joined the advancements?

7. Explain the functionality and purpose of n-gram models in language modelling, describing their role in early language processing.

8. Elaborate on the concept of Generative Adversarial Networks (GANs). Highlight the roles of the generator and discriminator in the generation of realistic content.

9. How do diffusion models work in generative modelling and what benefits do they offer in terms of image and audio generation?

10. Describe the significance of transformer models and neural-based embedding models for text generation in the context of AI.

11. Elaborate on the key aspects of prompt engineering, focusing on the importance of task framing, controlling the length of output and controlling the tone or nature of the output.

12. Explain the concept of prompt priming and how it differs from direct questioning. Describe the same with a suitable example.

13. What is the Chain-of-Thought (CoT) prompting approach and how does it contribute to improving large language models' performance on complex tasks?

14. Describe the few-shot learning technique in prompt engineering, emphasizing its role in teaching large language models specific tasks with a limited number of examples.

15. Explore the emerging trends and future directions in AI. Focus on technological advances, responsible AI practices and AI's expanded reach in various fields.

5.7.2 Subjective Question (5 Marks)

1. Explain the role of AI algorithms in identifying particles at the Large Hadron Collider (LHC) and the challenges it addresses.

2. How is AI utilized in astrophysics and space research, particularly in analyzing images of galaxies and simulating celestial phenomena?

3. In what ways does AI enhance efficiency and analysis in experimental chemistry, and what benefits does it bring to the discovery of novel compounds?

4. Describe the use cases of AI in biology, including predicting protein structures and tracking neural activity in the brain.

5. What is Prompt Engineering, and why is it considered a crucial skill for unlocking the true potential of ChatGPT?

6. How can the 'Explain to a 4th Grader' technique be effectively employed in prompt engineering, especially when dealing with complex concepts?

7. Discuss the role-play technique in prompt engineering and provide examples of scenarios where role-playing can be beneficial.

5.7.3 Objective Question (2 Marks)

1. Why GAN is called an adversarial network?

2. What are the different n-gram models?

3. What are some of the specialized hardware for AI?

4. What is Prompt Engineering?

5. What are the two benefits of using well-formed prompts?

6. Explain Aspect Based Sentiment Analysis (ABSA) with a suitable example.

7. Briefly describe, what you understand by machine translation.

8. What is causal AI?

5.7.4 Multiple-choice Questions (1 Mark)

1. How would you define generative AI in simple terms?
   (a) A type of AI used for deep learning
   (b) AI focused on discrimination tasks
   (c) AI that generates new content
   (d) AI specializing in computer vision

2. What is the key characteristic of discriminative models in machine learning?
   (a) They focus on generating new data
   (b) They predict outputs given inputs
   (c) They work only with images
   (d) They ignore input data

3. Which model introduced in 2014 revolutionized the field by using two neural networks against each other?
   (a) PixelRNN
   (b) Variational Autoencoders
   (c) Generative Adversarial Networks (GANs)
   (d) Autoencoders

4. What is the primary capability of Diffusion Models in generative modelling?
   (a) Predicting climate patterns
   (b) Denoising and generating realistic samples
   (c) Enhancing data analysis
   (d) Simulating astrophysical phenomena

5. What is prompt engineering?
   (a) Traditional software development
   (b) Art of asking questions to ChatGPT
   (c) Building physical structures
   (d) Writing computer code

6. In prompt engineering, what does 'Task Framing' involve?
   (a) Crafting a picture for the wall
   (b) Clearly expressing desired tasks in natural language
   (c) Creating a new programming language
   (d) Designing a piece of furniture

7. What is the purpose of the 'Explain to a 4th Grader' technique in prompt engineering?
   (a) Simplifying complex concepts for better understanding
   (b) Training 4th graders in advanced topics
   (c) Developing new educational materials
   (d) Testing knowledge of 4th-grade students

8. What is Chain-of-Thought (CoT) prompting in AI?
   (a) Sequentially chaining tasks in AI development
   (b) A method for improving large language models' performance
   (c) Creating a chain of AI-powered devices
   (d) Training AI models with thought processes

9. What is few-shot learning in prompt engineering?
   (a) Learning with very few examples
   (b) Intensive training with numerous examples
   (c) Simultaneous learning of multiple tasks
   (d) Learning without any examples

5.8 ANSWER KEYS

5.8.1 Multiple-choice Questions

1. (c) **2.** (b) **3.** (c) **4.** (b) **5.** (b) **6.** (b) **7.** (a) **8.** (b)
9. (a)

Model Question Paper

<< **NAME OF THE INSTITUTE** >>

DEPARTMENT OF << **XXX** >>

Final Semester Examination, November 2024

SUBJECT NAME: AI for Everyone

SUBJECT CODE: <XXX>

TIME: 3 HOUR   FULL MARKS: 80

PART A (Multiple Choice Type Questions)

1. **Answer any <u>ten questions</u>.** $10 \times 1 = 10$

   i. What significant event marked the formal emergence of the field of artificial intelligence?

      (a) The invention of the computer

      (b) The Dartmouth Conference

      (c) The development of LISP

      (d) The creation of the first AI program

   ii. Which type of machine learning model focuses on predicting outputs given inputs and does not generate new data points?

      (a) Generative models

      (b) Discriminative models

      (c) Variational autoencoders

      (d) PixelRNN

   iii. What should be an ideal first step in detecting bias in AI systems?

      (a) Check fairness metrics

      (b) Visual inspection

      (c) Review the data

      (d) Compare outcomes

iv. How does AI contribute to algorithmic trading?

(a) By slowing down trade execution

(b) By ignoring historical data

(c) By optimizing trade execution speed and analyzing vast amounts of historical data

(d) By minimizing the role of machine learning algorithms

v. How do e-commerce platforms utilize AI algorithms?

(a) To reduce customer engagement

(b) To limit personalization

(c) To predict consumer behaviour and offer personalized recommendations

(d) To ignore market trends

vi. Which government agency provided significant support for AI research during the formative years of the field?

(a) NSA

(b) DARPA

(c) CDC

(d) FDA?

vii. What is the potential downside of AI in terms of social inequality?

(a) It may increase access to education and healthcare.

(b) It can automate many tasks, leading to job displacement.

(c) It promotes diversity and inclusivity.

(d) It eliminates biases in decision-making

viii. What is one of the key reasons for the strong adoption of AI in the healthcare sector?

(a) Limited availability of medical data

(b) Low complexity of healthcare data

(c) High variability and complexity of medical data

(d) Minimal need for automation in medical analysis

ix. What is the purpose of prompt priming in ChatGPT?

(a) To directly ask questions

(b) To provide a step-by-step approach

(c) To generate creative content

(d) To optimize performance on complex tasks

x. What does the term 'accountability' signify in the context of AI systems?

   (a) Legal consequences

   (b) Ethical and moral expectations

   (c) Adherence to regulatory frameworks

   (d) Anticipation and corrective measures for decisions and actions

xi. What technological advancement laid the groundwork for AI research during World War II?

   (a) The invention of the smartphone

   (b) The development of quantum computing

   (c) The theoretical foundations of computing machines

   (d) The creation of social media platforms

Part B (Short Answer Type Questions)

Answer any <u>five questions</u> 5 × 5 = 25

2. Why has there been a larger adoption of AI in the healthcare industry? Explain the role of AI in early disease detection and prevention.

3. Explain, with example, the concept of fuzzy set. Compare fuzzy set with rough set.

4. Write the difference between:

   a. Supervised vs. unsupervised learning

   b. Transparency vs. Accountability in AI systems

5. Write short notes on:

   a. Object detection in computer vision

   b. N-gram models

6. What is machine translation? Explain the different approaches for machine translation.

7. Explain, with motivating examples, how prompt engineering can be used for improving efficiency in software development.

Part C (Long Answer Type Questions)

Answer any <u>three questions</u> 3 × 15 = 45

8. a) Explain, with a motivating example, bias and fairness in an AI system.   6

   b) Explain the different types of bias. Can bias be detected?   9

9. a) Explain few use cases related to financial risk management using AI. **6**

b) How can AI-enabled chatbots and virtual assistants used in different industries for achieving significant performance gain. **4**

c) Highlight some of the challenges of using AI in industry applications. **5**

10. a) Highlight the different versions of the GPT model from OpenAI elaborating the features brought forward in each version. **6**

b) Explain Chain of Thoughts technique of prompt engineering with a step-by-stem example. **4**

c) Compare the discriminative and generative models of AI. **5**

11. a) Explain the basic concept behind a recommendation system. What are the major categories of recommendation which are used in industry applications. **5**

b) Highlight some salient factors which has led to rapid adoption of deep learning in the last couple of decades. **6**

c) Outline, with real-life examples, how AI can promote equality. **4**

Index

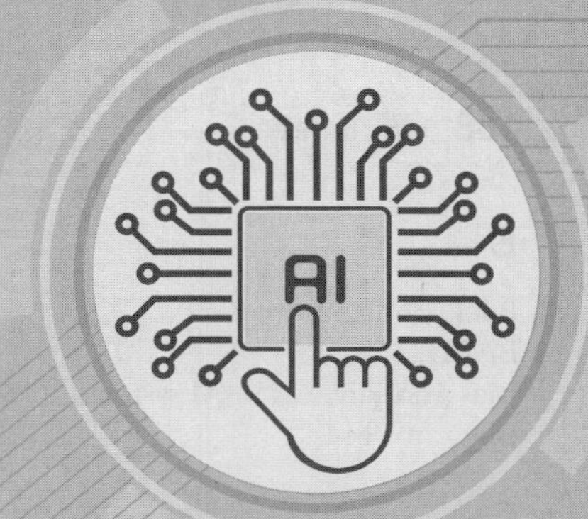

I

J

K

L

M

N

O

P

Q

Credits

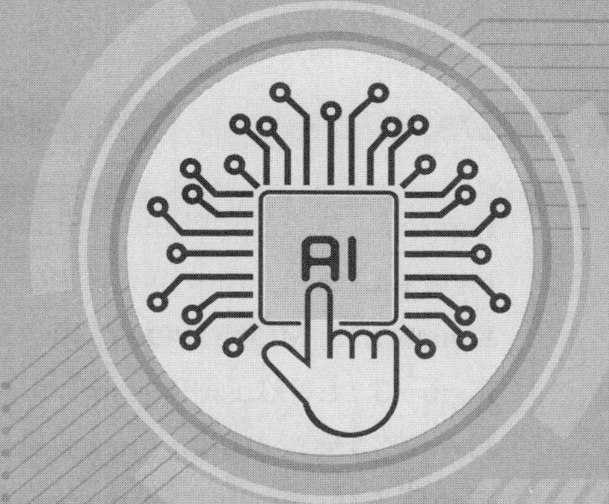

Figure 1.1: Everett Collection / Shutterstock

Figure 1.2: INTERFOTO / Personalities, INTERFOTO / Alamy

Figure 1.6: Zapp2Photo / Shutterstock

Figure 1.7: Zapp2Photo / Shutterstock

Figure 1.8: Sorn340 Studio Images / Shutterstock

Figure 1.9: Stock-Asso / Shutterstock

Figure 1.10: Ekkasit Keatsirikul / 123RF

Figure 1.11: metamorworks / Shutterstock

Figure 1.12: Suri_Studio / Shutterstock

Figure 2.1: Vlad Kochelaevskiy / Shutterstock

Figure 2.2: Zapp2Photo / Shutterstock

Figure 2.3: Martial Red / Shutterstock

Figure 2.4: Google Translate, Google and the Google logo are trademarks of Google LLC.

Figure 2.5: OpenAI

Figure 2.10: Nenov Brothers Images / Shutterstock

Figure 2.12: LDarin / Shutterstock

Figure 2.18: Richard Goldberg / Shutterstock

Figure 2.19: Anton Gvozdikov / Shutterstock

Figure 3.2: elnur / 123RF

Figure 3.3: Gorodenkoff / Shutterstock

Figure 3.4: SOMKID THONGDEE / Shutterstock

Figure 3.5: metamorworks / Shutterstock

Figure 3.7: MONOPOLY919 / Shutterstock

Figure 3.8: peshkov / 123RF

Figure 3.9: whiteMocca / Shutterstock

Figure 3.10: phonlamaiphoto / 123RF

Figure 3.11: Zapp2Photo / Shutterstock

Figure 3.12: ekkasit919 / 123RF

Figure 3.13: MONOPOLY919 / Shutterstock

Figure 3.14: Ekkasit Keatsirikul / 123RF

Figure 3.15: FeelGoodLuck / Shutterstock

Figure 3.16: Suwin / Shutterstock

Figure 3.17: kung_tom / Shutterstock

Figure 3.18: Visual Generation / Shutterstock

Figure 3.19: Stock-Asso / Shutterstock

Figure 3.20: Pavel Vinnik / Shutterstock

Figure 3.21: Eric Eric / 123RF

Figure 4.1 (Left): Noel Powell / Shutterstock

Figure 4.1 (Right): sirtravelalot / Shutterstock

Figure 5.1: nikonka1 / Shutterstock

Figure 5.4: Jan Martin Will / Shutterstock

Figure 5.5: Shutterstock.AI / Shutterstock

Figure 5.6: Google Books Ngram Viewer, http://books.google.com/ngrams

Figure 5.9: OpenAI

Figure 5.10: OpenAI

Figure 5.11: Google Gemini, Google and the Google logo are trademarks of Google LLC.

Figure 5.12: Google Gemini, Google and the Google logo are trademarks of Google LLC.

Figure 5.13: Google Trends, Google and the Google logo are trademarks of Google LLC.

Figure 5.14: Google Trends, Google and the Google logo are trademarks of Google LLC.

Figure 5.15(a): OpenAI

Figure 5.15(b): OpenAI

Figure 5.16: OpenAI

Figure 5.17: OpenAI

Figure 5.18: OpenAI

Figure 5.19: OpenAI